S0-BMG-621

Electron Microscopy
of Enzymes
Principles and Methods

Electron Microscopy of Enzymes
Principles and Methods

VOLUME I

Edited by

M. A. HAYAT

Associate Professor of Biology
Newark State College
Union, New Jersey

VNR VAN NOSTRAND REINHOLD COMPANY
New York   Cincinnati   Toronto   London   Melbourne

QP
601
.H334
V. 1

Van Nostrand Reinhold Company Regional Offices:
New York   Cincinnati   Chicago   Millbrae   Dallas

Van Nostrand Reinhold Company International Offices:
London   Toronto   Melbourne

Copyright © 1973 by Litton Educational Publishing, Inc.

Library of Congress Catalog Card Number: 73-4844
ISBN: 0-442-25676-0

All rights reserved. No part of this work covered by the copyright
hereon may be reproduced or used in any form or by any means—
graphic, electronic, or mechanical, including photocopying, recording, taping, or information storage and retrieval systems—without
permission of the publisher.

Manufactured in the United States of America

Published by Van Nostrand Reinhold Company
450 West 33rd Street, New York, N.Y. 10001

Published simultaneously in Canada by Van Nostrand Reinhold Ltd.

15 14 13 12 11 10 9 8 7 6 5 4 3 2 1

Library of Congress Cataloging in Publication Data

Hayat, M   A
    Electron microscopy of enzymes.

    Includes bibliographies.
    1. Enzymes. 2. Electron microscope—Technique.
I. Title. [DNLM: 1. Enzymes. 2. Microscopy,
Electron. QU 135 H415e 1973]
QP601.H334   574.1'925'028   73-4844
ISBN 0-442-25676-0 (v.1)

Library
UNIVERSITY OF MIAMI

JMC 2-27-75

It is a pleasure to dedicate this volume to

Russell J. Barrnett and A. G. Everson Pearse

Preface

Tested and fairly reliable electron cytochemical methods are now available. Although these methods are being modified and improved rapidly, a comprehensive treatise on these methods is appropriate. This volume is presented since no book containing procedures for the preparation of biological specimens for electron cytochemistry is available. The primary objective of this book is to provide the reader with a detailed description of the methodology employed to localize enzymatic activity at the subcellular level. The procedures are presented in a stepwise form, so that they can be carried out with ease and without outside help.

The topics were carefully selected and written by competent investigators from several countries who have pioneered in their respective fields. The contributors have reviewed their subjects in a fair and factual manner. Alternative procedures, points of disagreement, and potential research areas are stated. The contributors have focused on those aspects of electron cytochemical methods which have contributed much to the localization of enzymatic activity in the past and are likely to be in the vanguard of improvements to come. Omissions should be ascribed to the limitations of my own particular view, and should not be construed as a judgment of any of the contributors. An introductory chapter on fixation with aldehydes is included because this information is fundamentally important to anyone contemplating work in cytochemistry. This chapter presents guidelines for a newcomer to electron cytochemistry.

The reader should find this book an excellent reference, for it summarizes the major procedures which have evolved within the past two decades. An exhaustive list of references with complete titles is provided for each chapter, as are full author and subject indices.

The validity of techniques demonstrating true sites of enzymatic activity has been repeatedly questioned. This controversy has reached almost colossal proportions. The well-known vagaries of electron cytochemical methods involving the use of lead salts and the resulting uncertainty have already led to the development of alternative techniques. The discovery of technical problems, especially the specificity of substrates, need not lead to disillusionment and pessimism, but to renewed and vigorous effort in designing better and more effective procedures. It is my hope that this book will contribute significantly to the development and standardization of methodology.

In spite of my best efforts, it was not possible to present various groups of enzymes in a logical sequence. It was considered undesirable to delay the publication of chapters which were completed first. Since it is almost impossible to present techniques employed for the localization of enzymes in a single volume, the preparation of the second and third volumes under the same main title is in progress.

The problem with any book which attempts to describe events in a rapidly progressing technical field is that it needs updating very soon after its publication. This problem will be met by keeping the reader abreast of recent developments in subsequent editions.

I am deeply indebted to Professor Russell J. Barrnett for his invaluable suggestions during the preparation of this volume. Also, I am grateful to Drs. Burton L. Baker, Edward Essner, F. Hajós, and Theodore K. Shnitka for their help. I appreciate the gracious help given to me by Miss Gerda Zeiselmeier during the preparation of this volume.

M. A. Hayat

Contents

ix

2 PHOSPHATASES
Edward Essner

3 GLYCOSIDASES β-GLUCOSIDASES, β-GLUCURONIDASE
I.D. Bowen

4 GLYCOSIDASES N-ACETYL-β-GLUCOSAMINIDASE
D. Pugh

5 GLUTAMATE OXALACETATE TRANSAMINASE
Sin Hang Lee

6 MYROSINASE IN CRUCIFEROUS PLANTS
Tor-Henning Iversen

7 ENZYME IMMUNOCYTOCHEMISTRY
Ludwig A. Sternberger

Contents of

Electron Microscopy
of Enzymes
Principles and Methods

1

Specimen Preparation

M. A. HAYAT

Department of Biology,
Newark State College,
Union, New Jersey

INTRODUCTION

Because of its increased direct magnification, an electron microscope is a powerful tool for the examination of cell structure and elucidation of its function. The examination of cellular structures has become more meaningful since the introduction of high resolution electron microscopes which, under favorable conditions, have the capability of a point to point resolution of better than 3 Å. It should be remembered that, in terms of its importance, there is no substitute for direct visualization of cellular materials.

Electron cytochemistry of enzymes offers the opportunity of escaping from the limitations of the light microscope in terms of accuracy of localization and specificity. To examine enzymatic activity without at least adequate preservation of cellular structure is of little use. The exact location of an enzyme can be determined only with reference to the preserved cellular organelles. When several enzymes possess overlapping substrate specificities, the activity of each enzyme can be identified by relating it to a specific identifiable cell organelle. This is the primary reason for expending so much effort to develop techniques which would preserve not only enzymatic activity but also cellular fine structure.

Although fresh tissues have been utilized for the preservation of enzymatic activity, they are not ideally suited for routine ultrastructural localization of enzymes. The difficulties encountered are: inadequate preservation of fine structure, diffusion and displacement of enzymes into the surrounding medium, permeability barrier maintained by membranes, and difficulty in handling (Shnitka and Seligman, 1971).

In general, tissues fixed with glutaraldehyde require shorter incubation periods than those required for fresh tissues. Relatively slow penetration of the reaction medium into fresh animal (Ericsson and Trump, 1965) and plant (Sexton *et al.*, 1971) tissues has been reported. The penetration of the reaction medium into fresh tissues is bound to be impeded by the natural permeability barriers of the intact cells. In cultured cells, however, penetration does not constitute a significant problem (Halperin, 1969).

At present, aldehydes, osmium tetroxide, and potassium permanganate are the reagents of choice to preserve cellular fine structure for conventional electron microscopy. Because osmium tetroxide and potassium permanganate are heavy metal oxides with an oxidizing capacity of destroying enzymatic activity, their use in enzyme cytochemistry is limited; whereas many cytochemical reactions can be performed on tissues after fixation with an aldehyde. Also, relatively large size ($\sim$ 8 mm) tissue blocks can be prefixed in aldehydes, especially with paraformaldehyde. Prior to further treatment, the prefixed blocks can be cut into smaller pieces with minimal mechanical damage, since aldehyde fixation imparts some degree of firmness to these blocks. Prefixation with an aldehyde thus lessens the distortions introduced by mincing fresh tissues prior to their incubation and postfixation with osmium tetroxide. For these and other reasons discussed later, aldehydes are used widely for the study of enzyme activity at the subcellular level.

Although formaldehyde and other aldehydes have been employed since 1951 (Seligman *et al.*, 1970) for the preservation of enzymatic activity at the light microscope level, their use in electron microscope cytochemistry is of rather recent development. One of the early attempts to employ aldehydes for enzyme localization at the subcellular level was made by Essner *et al.* (1958). They employed 4% formaldehyde containing 1% calcium chloride prior to incubation to preserve phosphate-ester splitting enzymes. In these and other studies at that time, although the preservation of fine structure was less than satisfactory, enzymatic activity was adequately preserved. A serious attempt to explore the usefulness of several mono- and dialdehydes as fixatives for the preservation of both enzymes and fine structure was made in 1963-64 by Sabatini, Barrnett, and associates. Since then, progress in developing better methods for the preservation of enzymatic activity at the subcellular level has been rapid.

At present, methods are available for satisfactory fine structural demonstration of enzymes, and fixation with aldehydes is a prerequisite for many of these methods. The difference in the reactivity of various aldehydes with proteins can be utilized to select an appropriate aldehyde for the preservation of a given enzyme. Also, the duration of fixation and concentration of the fixative can be varied. Thus, it is possible to demonstrate a wide spectrum of enzymes ranging from highly resistant alkaline phosphatases to labile succinic dehydrogenase by choosing conditions that are appropriate for the adequate preservation of both enzyme activity and fine structure.

REACTION OF ALDEHYDES WITH PROTEINS

Although aldehydes preserve enzymatic activity better than other fixatives, they invariably alter protein structure. However, the possibility that some proteins may escape modification during fixation should not be overlooked. Aldehyde-mediated alterations result in the inactivation of enzymes because many enzymatic activities depend upon tertiary protein structure. This type of enzymatic inactivation is caused by direct effects on the enzyme site. Aldehydes also affect membrane permeability and the compartmentalization and binding of ions. These alterations may influence enzymatic activities. Aldehydes may also block substrate and carrier sites. Among commonly employed aldehydes, glutaraldehyde causes the least conformational changes in protein, although some structural modifications in the protein molecule do occur, especially in the α-helix. It has been shown that incubation of bovine hepatic fructose-1,6-diphosphatase with glutaraldehyde alters both the catalytic and allosteric properties of the enzyme (Aloyo et al., 1972).

The effectiveness of aldehydes in the partial preservation of enzymatic activity is due to the fact that they are non-coagulant fixatives. Thus they transform proteins into a transparent rather than an opaque gel, and stabilize proteins, including enzymes, without much structural distortion of the original state. Aldehydes cause very little dissociation of protein from water, and proteins retain at least some of their reactive groups. Another desirable effect of aldehydes is that they can render certain proteins noncoagulable by subsequently used coagulant fixatives and dehydration solvents. The importance of this becomes apparent when the fixed tissue is dehydrated with ethanol. Biochemical studies indicate that glutaraldehyde-treated enzymes are highly resistant to denaturation. Glutaraldehyde-stabilized enzymes may be considered to be desensitized. Aldehydes are also additive fixatives, for they become chemically part of the proteins they stabilize.

The major groups in proteins that may react with aldehydes are given below.

Arginine

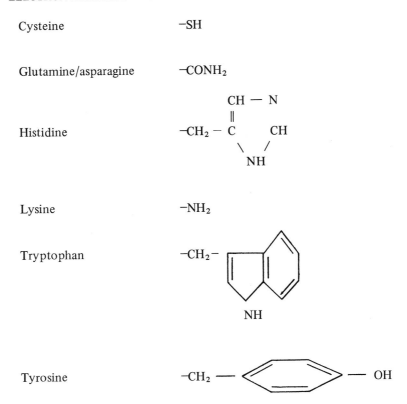

Cysteine $-SH$

Glutamine/asparagine $-CONH_2$

Histidine

Lysine $-NH_2$

Tryptophan

Tyrosine

Various aldehydes differ in the rate and extent of their reaction with the above reactive groups in proteins. The extent of protein reaction is influenced by several factors, including the duration and mode of fixation, pH, temperature, conformation of the protein molecule, and the type and concentration of the aldehyde employed. The variations in the interaction between proteins and aldehydes are considered to be correlated with the effects of the latter on enzyme activity. For further information on the mechanism involved in the interaction between various aldehydes and proteins, the reader is referred to Hayat (1970).

Glutaraldehyde has been employed to covalently link enzymes to carriers in immunocytochemistry (Avrameas and Ternynck, 1969). The tertiary structure of the enzyme molecule seems to remain largely intact in the biogel-enzyme conjugate. In addition, a high proportion ($\sim$ 80%) of antigenic activity is retained. By using glutaraldehyde, for instance, acid phosphatase has been linked to polyacrylamide beads; thus, more than 55% of the enzymatic activity was retained (Weston and Avrameas, 1971). Water insolubilization of enzymes with polyaldehydes derived from cross-linked polyacryloylaminoacetaldehyde dimethylacetal has also been reported (Epton et al., 1971). (See Sternberger, in this volume.)

EFFECTIVENESS OF VARIOUS ALDEHYDES

As stated above, various aldehydes differ in their ability to preserve enzymatic activity and morphology of cell components (Table 1-1). It is emphasized that Table 1-1 is only for general use, for there are many exceptions. The activity of a given enzyme may be best localized by a certain aldehyde which may not be suitable for the preservation of activity of most other enzymes. In general, hydrolytic enzymes (e.g., phosphatases) are more resistant to aldehyde fixation than are oxidative enzymes (e.g., succinic dehydrogenase). Thus, in order to obtain an appreciable degree of activity of the latter enzymes, the excellence in the preservation of ultrastructure may have to be sacrificed.

Even the same type of enzyme in a similar organ may differ in its resistance to aldehyde fixation, depending upon the species of the animal. Glutaraldehyde fixation, for instance, caused considerable reduction in rat hepatic glucose-6-phosphatase even after immersion for 10 min, and the reaction was almost abolished after immersion for 30 min (Kanamura, 1970). On the other hand, the reaction in mice was only moderately reduced after immersion for 10 min and considerably reduced after immersion for 30 min. As expected, it was also found that the reaction was reduced more rapidly in fasted animals. The difference in resistance to aldehyde fixation between mouse and rat hepatic glucose-6-phosphatase is probably mainly due to the difference in the properties of the enzyme and only slightly due to the difference in the quantity of the enzyme.

Table 1-1 Effects of Aldehydes on the Preservation of Fine Structure and Enzymatic Activity

	Preservation	
	Ultrastructural morphology	*Enzymatic activity*
Acetaldehyde	poor	good
Acrolein	good-excellent	poor
Crotonaldehyde	fair-good	good
Glutaraldehyde	excellent	good
Glyoxal	fair-good	good
Hydroxyadipaldehyde	poor-fair	excellent
Malonaldehyde	fair	good
Methacrolein	good	good
Paraformaldehyde	fair	excellent
Pyruvic aldehyde	poor	fair-good
Succinaldehyde	fair	good

It is pointed out that various aldehydes differ in their effect on the appearance of different organelles. Endoplasmic reticulum in chicken heart cells, for example, consists of small and branched cisternae following treatment with glutaraldehyde and acrolein, while formaldehyde causes a measurable increase in the diameter of the cisternae. The quality of organelle preservation following fixation with various aldehydes is shown in Table 1–2.

Alkaline phosphatases and ATPases can be localized after fixation with all the aldehydes listed in Table 1-1 except acrolein. The activity of acid phosphatase in rat and slug liver (Bowen, 1971), in rat lung (Etherton and Botham, 1970), and in the descending part of the proximal tubule of rat kidney (McDowell, 1969), β-glucuronidase in rat and slug liver (Bowen, 1971), and glucose-6-phosphatase in hepatocytes (Leskes et al., 1971) and in rat kidney (Goldfischer et al., 1963) can be preserved with glutaraldehyde. Studies of the inhibitory effects of three aldehydes on glucose-6-phosphatase activity in fresh-frozen sections of rat liver fixed by immersion indicated that glutaraldehyde (3%) caused a marked reduction in enzyme activity after a 1 min fixation (Ericsson, 1966). A corresponding reduction in the enzyme activity occurred following fixation for 5 min in paraformaldehyde (2%) and 10 min in hydroxyadipaldehyde (12.5%). However, maximum preservation of the enzyme activity was obtained after the liver was fixed with glutaraldehyde by vascular perfusion of a short duration (1 to 2 min).

Comparative studies of the effect of eight aldehydes on the preservation of glucose-6-phosphatase in cultured chicken heart cells indicate that glutaraldehyde and hydroxyadipaldehyde permit excellent demonstration, methacrolein,

Table 1-2 Effect of Aldehydes on Organelle Preservation

	Ground cytoplasm	Endoplasmic reticulum	Perinuclear space	Mitochondria	Golgi complex
Glutaraldehyde	+ + + +	+ + + +	+ + + +	+ + + +	+ + + +
Acrolein	+ + + +	+ + + +	+ + + +	+ + + +	+ + +
Paraformaldehyde	+ +	+ + +	+ +	+ + +	+ +
Hydroxyadipaldehyde	+	+	+ +	+ +	
Methacrolein	+	+	+		+ +
Crotonaldehyde	+ + +			+	+
Glyoxal		+		+	+ +
Acetaldehyde		+	+ + +		

From M. Hündgen et al. (1971). Used by permission.

glyoxal, and acetaldehyde permit limited demonstration, and acrolein and para-formaldehyde permit almost no demonstration of the enzymatic activity (Schä-fer and Hündgen, 1971). Hydroxyadipaldehyde has been employed for the preservation of succinate dehydrogenase activity in the mitochondria of cardiac and skeletal muscles of the mouse (Makita and Sandborn, 1971).

Glutaraldehyde is effective in preserving peroxidases in leaves (Frederick and Newcomb, 1969). This dialdehyde (8%) has also been used to selectively destroy the hemagglutinin of certain strains of influenza virus while preserving 50 to 60% of the neuraminidase activity (Blough, 1966). There is, however, some evidence which indicates that labile enzymes such as cytochrome oxidase and succinic dehydrogenase are best preserved by hydroxyadipaldehyde. Succinic dehydro-genase activity has also been preserved by crotonaldehyde (Ogawa and Barrnett, 1965).

Quantitative studies of the effect of glutaraldehyde and formaldehyde on several enzymes (β-galactosidase, N-acetyl-β-glucosaminidase, acid phosphatase, β-glucuronidase, catalse, and various dehydrogenases and amino-transferases) indicated that the loss of enzymatic activity with the former was approximately twice that following treatment with the latter (Janigan, 1964 & 1965; Hopwood, 1967a; Anderson, 1967). Acid phosphatase activity in the nuclei of monolayers of Hela cells was preserved much better with formaldehyde than with glutaralde-hyde (Soriano and Love, 1971). It has been shown that glutaraldehyde fixation is more inhibitory than formaldehyde for nucleoside phosphatase and NDPase activity in the rat anterior pituitary gland (Pelletier and Novikoff, 1972). The activity of ATPase in the rat brain was preserved better when the tissue was prefixed with formaldehyde than with glutaraldehyde (Torack, 1965).

That more than 70% of the Na-K-ATPase activity in avian salt gland remains after paraformaldehyde (3%) fixation has been demonstrated (Ernst and Phil-pott, 1970; Ernst, 1972). Oxidative activities in root cells were best preserved with 4% paraformaldehyde (Nir and Seligman, 1971). Formaldehyde-calcium chloride was recently employed for the localization of N-acetyl-β-glucosamini-dase activity in rat tissues (Pugh, 1972).

Glutaraldehyde has been reported to be approximately twenty times more potent than formaldehyde in the inhibition of extra ATPase activity in isolated vesicles of the sarcoplasmic reticulum of rabbit skeletal muscle (Sommer and Hasselback, 1967). Biochemical studies have shown that aryl sulfatase activity is inhibited to some degree by glutaraldehyde (Hopsu-Havu et al., 1967). Glutar-aldehyde-fixed monkey brain showed ~20% residual succinate dehydrogenase, 40% acetate dehydrogenase, monoamine oxidase, phosphatase, and ATPase activity compared to the activity observed in fresh-frozen sections (Manocha, 1970). It should be noted that even with the latter technique a certain amount of enzyme activity is lost.

The above-mentioned and other studies indicate that, in general, glutaralde-hyde permits only moderate preservation of enzyme activity in comparison to formaldehyde. This difference is probably due in part to slow penetration and rapid fixation by glutaraldehyde in contrast to formaldehyde, which diffuses rapidly but fixes slowly. This difference in the rate of penetration may be due to the difference in the size of the molecule of these two aldehydes. In this connection, the difference in the rate and/or nature of denaturation of proteins by these two aldehydes is also important.

It is generally thought that the quality of preservation of the fine structure is proportional to the number of cross-links introduced by the aldehyde with proteins, while the preservation of enzyme activity is inversely proportional to the number and speed of cross-links introduced. This is the reason why glutaraldehyde is not as efficient as formaldehyde or hydroxyadipaldehyde is in preserving enzymatic activity.

Formaldehyde generated from paraformaldehyde is used widely in electron cytochemistry. Paraformaldehyde is preferred over formaldehyde (as it is commercially available) because the latter contains undetermined amounts of impurities such as methanol and formic acid. Paraformaldehyde, for example, has been most effective in the localization of mitochondrial ATPase in cultured chicken cardiac myoblasts (Ahrens and Weissenfels, 1969) and cytochrome oxidase activity in hepatic and cardiac cells (Seligman *et al.*, 1970). Paraformaldehyde polymer is slowly and partially soluble in water at 60°C. The addition of a few drops of 1 N NaOH accelerates and completes dissolution of the powder. Water produces a partial hydrolysis of the polymer, which is completed by the presence of a small amount of alkali. Solutions should be prepared immediately prior to use.

It must be pointed out that cross-linking does not always inhibit enzymatic activity and that even the preservation of a small percentage of residual enzymatic activity can be sufficient for precise localization. A recovery of 12% of acid phosphatase activity, for example, is sufficient to be visualized by light and electron microscopy. It is possible that even only some of the amino groups present in the enzyme molecule can maintain the activity of the enzyme. Thus, glutaraldehyde is a very useful fixative for enzyme localization. In fact, by a suitable choice of conditions, even the most labile enzymes can be detected with cross-linking aldehydes such as glutaraldehyde. Controlled application of glutaraldehyde should, indeed, be very useful in preserving enzymatic and antigenic activities. Since the best preservation of fine structure is obtained with glutaraldehyde, it has become the most widely employed aldehyde fixative for electron microscopy. Table 1-3 presents the comparative inhibition of various enzymes after treatment with glutaraldehyde.

The ideal approach apparently is to develop a formulation which will give satisfactory preservation of both enzymatic activity and fine structure. The development of such a formulation would not be very difficult, since sufficient biochemical information regarding the differences in the reactivities of various aldehydes with proteins is available (Hayat, 1970). By employing appropriate

Table 1-3. Comparative Inhibition of Enzymes in Various States by Glutaraldehyde.

Enzyme	State	Reaction time	%Activity remaining	% Glutaraldehyde (g/100 ml)	Temp. (°C)	pH	Authors
Acid phosphatse	Tissue blocks	1hr	40	4	4	7.3	Anderson, 1967
		2hr	28	4	4	7.2	Hopwood, 1967a
		6hr	19–14	4	2	7.2	Janigan, 1965
		18hr	15	4	4	7.2	Hopwood, 1967a
		24hr	12	4	2	7.2	Janigan, 1965
N-acetyl-β-glucosaminidase	Perfused tissue	5min	12	1.5	4–20	7.4	Arborgh et al., 1971
Alanine aminotransferase	Tissue blocks	18hr	54[a]	4	2	7.2	Janigan, 1964
Aryl sulfatase	Tissue blocks	1hr	50	4	4	7.3	Anderson, 1967
Aspartate aminotransferase	Perfused tissue	5min	30	1.5	4–20	7.4	Arborgh et al., 1971
	Homogenate film	5min	30–32	1	4	7.2	Papadimitrio and van Duijn, 1970
ATPase, Na- & K-activated	Homogenate	40–60 min	0	0.5	4	7.3	Ernst and Philpott, 1970
ATPase, Mg-activated	Homogenate	40–60 min	15	0.5	4	7.3	Ernst and Philpott, 1970
Carboxypeptidase	Crystalline	1hr	40	1	23	7.5	Quiocho and Richards, 1966
Catalase	Crystalline	1hr	12	4	23	7.2	Schejter and Bareli, 1970
	Tissue blocks	2hr	12.5	4	4	7.2	Hopwood, 1967a
		18hr	6.6	4	4	7.2	Hopwood, 1967a
Cholinesterase	Tissue blocks	1hr	75	4	4	7.2	Anderson, 1967
α-Chymotrypsin	Crystalline	1hr	0.4–1.2	2.3	0	6.2	Jansen et al., 1971
iso-Citric dehydrogenase	Tissue blocks	1hr	20	4	4	7.3	Anderson, 1967
Creatinine phosphokinase	Tissue blocks	1hr	12	4	4	7.3	Anderson, 1967
β-Galactosidase	Tissue blocks	7hr	43	4	2	7.2	Janigan, 1964
		24hr	24	4	2	7.2	Janigan, 1964
β-Glucuronidase	Tissue blocks	2hr	38	4	4	7.2	Hopwood, 1967a
		7hr	24	4	4	7.2	Janigan, 1964
		18hr	15	4	2	7.2	Hopwood, 1967a
		24hr	12	4	4	7.2	Janigan, 1964
Glycogen phosphorylase b	Crystalline	10min	40	0.05–0.01	23	7.5	Wang and Tu, 1969
α-Hydroxybutyrate dehydrogenase	Tissue blocks	1hr	30	4	4	7.3	Anderson, 1967
Lactate dehydrogenase	Tissue block	1hr	13	4	4	7.3	Anderson, 1967
Ribonuclease	Crystalline	–	32	4	4	7.2	Sachs and Winn, 1970
Subtilisn	Crystalline	30–60 min	10–15	2	23	6–8	Ogata et al., 1968

Note: a = tissue washed in running water before assay. From D. Hopwood, *Histochem. J.* 4, 267, 1972. Used by permission.

mixtures of aldehydes, it should be possible to demonstrate cytochemically a wide spectrum of enzymes as well as satisfactory preservation of the fine structure. A mixture of paraformaldehyde and glutaraldehyde (Karnovsky, 1965), for instance, has been used for the localization of peroxidase (Cotran and Karnovsky, 1968). This mixture was recently employed to fix amoeba for the localization of AcPase and TPPase in the Golgi and endoplasmic reticulum (Wise and Flickinger, 1971) and heart for the localization of glutamic oxalacetic transaminase (Lee et al., 1971).

LIMITATIONS OF ALDEHYDES

Caution is warranted in the use of glutaraldehyde, for evidence has been presented indicating the activation of a nuclear acid phosphatase enzymatic activity as a result of treatment with this dialdehyde (De Jong et al., 1967). Another type of possible artifact is formed because of the ability of glutaraldehyde to immobilize metabolites such as amino acids. Glutaraldehyde is thus capable of binding, for instance, free amino acids to tissue constituents. This becomes significant in autoradiographic studies where amino acids labeled with radioisotopes are employed. Peters and Ashley (1967) have described a possible artifact caused by glutaraldehyde in the study of protein formation in the presence of labeled amino acids. On the other hand, this binding capability of glutaraldehyde can be utilized advantageously to immobilize diffusible compounds, including enzymes containing amino acid groups. In this connection, it is known that this dialdehyde penetrates into the tissue before labeled amino acids are lost from the cell by diffusion.

Studies of the effect of glutaraldehyde on the levels of biochemical constituents in rat brain indicated that, following perfusion fixation, significant increases in the level of brain glutamic acid, alanine, valine, isoleucine, leucine, and tyrosine occurred compared with unfixed controls (Davis and Himwich, 1971). This is expected since, through cross-linking, glutaraldehyde binds amino acids initially and then larger amounts are allowed to be extracted for the analyses. The increase in the amino acid level is also attributed to the shrinkage of the tissue because the level of amino acids is based on wet tissue weight.

The contraction of extracellular space in the nervous tissue during glutaraldehyde fixation is a well-known phenomenon. Van Harreveld and Khattab (1968) have demonstrated that perfusion of cortex with glutaraldehyde caused an increase in the impedance of the tissue, an accumulation of chloride into cellular elements, and a contraction of extracellular space. This transport of extracellular material is thought to be a consequence of an increase in the sodium permeability of the plasma membrane of the cellular elements which take up chloride and water during fixation with glutaraldehyde. Recent studies by Van Harreveld and Fifkova (1972) indicated that glutamate is released from the intracellular into the extracellular compartment in thick retina during fixation with glutaraldehyde, and it was suggested that the action of glutamate on the plasma membrane is responsible, in part, for the contraction of extracellular space.

In addition, longer durations of fixation with glutaraldehyde may increase the production of artifactual myelin figures. Aldehyde fixation in the cold may result in the loss or rearrangement of microtubules and a more dispersed pattern of ribosomes.

Aldehydes are unable to impart sufficient contrast and density to the tissue, and have no staining action. In addition, they are incapable of rendering lipids insoluble in organic solvents. Accordingly, aldehyde-fixed tissues show cellular membranes as negative images. These limitations are overcome by postfixation with osmium tetroxide or potassium permanganate. Postfixation also stabilizes the fine structure already maintained by the aldehyde so that it can withstand embedding in resin. Also, postfixation with osmium tetroxide is an absolute requirement for electron cytochemical methods that are based on the "osmiophilic principle."

However, the use of postfixation is not always desirable. For a valid demonstration of acid phosphatase in nerve cell lipofuschin bodies, for instance, postfixation with osmium tetroxide should be avoided. Lipofuschin is osmiophilic, and thus it is difficult to separate lead phosphate precipitate from the reaction product of osmium tetroxide with the contents of lipofuschin granules (Brunk and Ericsson, 1972). In addition, longer periods of postfixation may be equivalent to an "acid rinse," which can dissolve the lead phosphate precipitate (Desmet, 1962; Reale and Luciano, 1964).

Aldehydes cause shrinkage of the cell, including the nucleus. Glutaraldehyde (4% at pH 7.4), for instance, causes a shrinkage of ~6% of rat liver in 18 hr at $4°C$ (Hopwood, 1967b), while 2% glutaraldehyde causes 6 to 10% shrinkage of calf erythrocytes in 10 min (Cartensen et al., 1971). After glutaraldehyde fixation, the gain in dry weight of plant tissues was reported to be ~1.1% of the fresh weight (Morré and Mollenhauer, 1969).

FACTORS AFFECTING THE PRESERVATION OF ENZYMATIC ACTIVITY

Satisfactory preservation of enzymatic activity is controlled by many factors, including the concentration and rate of penetration of the fixative solution, duration and mode of fixation, rate of penetration of the incubation medium, mode of incubation, temperature, pH, buffer, and specimen size. In addition, the duration elapsed between the death of the organism and processing of the tissue (i.e., postmortem interval) cannot be overlooked. These factors are elaborated upon below.

Mode of Fixation and Incubation

The available data indicate that the quality of preservation of the fine structure and enzymatic activity is affected not only by the characteristics of the fixative but also by the method of applying the fixative to the tissue. Under *in vivo* conditions, the rate and depth of fixation are increased. Fixation by immersion, on the other hand, may produce artifacts. In the tissues fixed by immersion, the total activity of blood and tissue enzyme is measured, which introduces a nonsystematic fault by the variability of the amount of blood remaining in the organs. Vascular perfusion, in contrast, removes blood almost completely from the organs. This has been confirmed by von Matt *et al.* (1971), who quantitatively determined rat liver acid phosphatase activity in perfused and decapitated animals. They found that the perfusion method significantly increased the reproducibility of the results. The nature and frequency of artifacts caused by immersion fixation are not fully known.

Although fixation by vascular perfusion is superior to fixation by immersion, the latter method is much more widely used. The reason for this is that not many tissues can be fixed by vascular perfusion—for instance, *in vitro* fixation becomes a necessity in the case of human tissues. In addition, perfusion fixation is rather expensive and elaborate, and a high degree of skill is required to carry it to successful completion. The success of the perfusion method depends, in part, upon the complete exclusion of blood tissue from the vascular system and the prevention of vasoconstriction (blocking or narrowing of the blood vessels).

Perfusion fixation has proved an effective approach to solving certain problems encountered during preserving enzymatic activity. A mild but effective fixation can be obtained by a short perfusion, which results in the preservation of a higher percentage of enzymatic activity. Longer durations of fixation tend to cause diffusion and inactivation of enzymes. Venous perfusion (Ericsson, 1966) and transparenchymal perfusion (Kanamura, 1971a) by glutaraldehyde have been employed for cytochemical demonstration of glucose-6-phosphatase activity in rat liver. As much as 70% of the total activity of this enzyme has been preserved after fixing the liver with 1% glutaraldehyde by vascular perfusion (Casanova *et al.,* 1972). A mixture of glutaraldehyde (1%) and paraformaldehyde (3.7%) was employed to perfuse rat heart for the localization of glutamic oxalacetic transaminase activity (Lee *et al.,* 1971).

As emphasized earlier in the case of immersion fixation, in general the concentration of the fixative, the duration of perfusion, and temperature (i.e., $4°C$) are critical factors for the optimal preservation of enzymatic activity. Even a slight change in these factors can adversely affect the results. This was demonstrated by Kanamura (1971a) for the localization of glucose-6-phosphatase activity in hepatocytes—no deposition of reaction product was found when the duration of transparenchymal perfusion was increased from 3 to 5 min. Other studies, however, indicate that glucose-6-phosphatase is not as sensitive as suggested by Kanamura (1971a). This enzyme can also be demonstrated by relatively long immersion fixation. It is admitted that at present the evidence is somewhat conflicting and that the reasons for the discrepancies are not entirely clear. In this connection, the mode of fixation and tissue type are important factors. For general purposes, perfusion with 1% glutaraldehyde for 2 min is recommended. Detailed procedures for the vascular perfusion of lung, heart, liver, brain, kidney, and muscle have been presented (Hayat, 1970; 1972).

Limited evidence is available to indicate that the addition of hydrogen peroxide to glutaraldehyde results in better preservation of both fine structure (Fig. 1-1) and enzymatic activity. Improvements in the preservation of fine structure in the specimens treated with hydrogen peroxide have been demonstrated (Peracchia et al., 1970; Peracchia and Mittler, 1972a and b). The effectiveness of hydrogen peroxide treatment has been presumed to depend upon the activity of reaction products of hydrogen peroxide and glutaraldehyde, possibly α-β-epoxyaldehydes. Recent studies on nucleoside diphosphatase and thiamine pyrophosphatase activities in hepatocytes and other cells of rat indicate that glutaraldehyde—hydrogen peroxide preserves enzymatic activity better than conventional glutaraldehyde (Goldfischer et al., 1971). Improvements in the tissue preservation as well as in the speed and depth of fixative penetration have also been reported in the same study. In addition, frozen sections of tissues fixed in the presence of hydrogen peroxide have been reported to be less friable and easier to cut.

Another approach is to incubate the tissue during fixation by either immersion or vascular perfusion. In this procedure, the substrate protects the active site of the enzyme by competing with the aldehyde. Inactivation of asparate aminotransferase, for example, was delayed by adding ketoglutarate to the aldehyde (Papadimitriou and van Duijn, 1970). Leskes et al. (1971) successfully incubated rat liver during fixation by perfusion via portal vein, and succeeded in recovering glucose-6-phosphatase activity approximately 80% of that present prior to fixation. The perfusion was carried out with glutaraldehyde (2%) and lasted for 3 to 5 min. The exposure of the tissue to the substrate during fixation did not contribute any detectable background precipitate. Kanamura (1971b) also indicated that, in the absence of the substrate from the fixation solution, the deposition of glucose-6-phosphatase activity in proximal convoluted tubule cells of rat kidney was diminished.

An additional approach is to incubate the tissue prior to fixation. It has been claimed that as long as the duration of incubation does not exceed 20 to 50 min, the intracellular components retain their accepted morphological appearances. Hajós and Kerpel-Fronius (1970) have demonstrated the preservation of succinic dehydrogenase activity in the mitochondria as well as in the ultrastructure of rat myocardium after an incubation of 45 min prior to fixation with 4% paraformaldehyde followed by 1% osmium tetroxide. Excellent preservation of succinic dehydrogenase activity and coupled succinate oxidation by incubation prior to fixation are shown in Figs 1-2 and 1-3.

It must be emphasized, however, that incubation without prior fixation may invite diffusion artifacts which would be difficult to interpret correctly. It is known, for instance, that liver esterase tends to diffuse from unfixed sections; the thinner the section, the more diffusion will occur (Barrow and Holt, 1971).

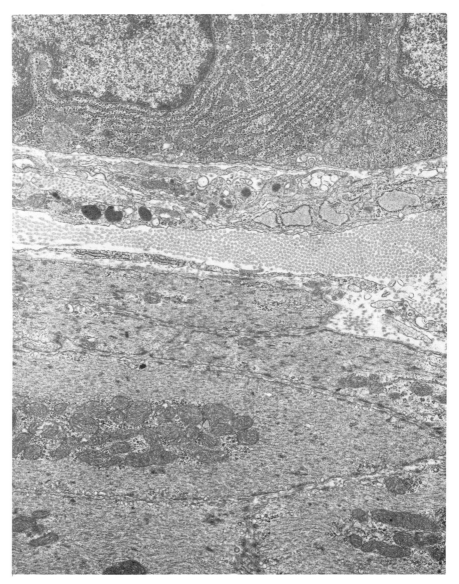

Fig. 1–1. Rat jejunum fixed with glutaraldehyde-H_2O_2. Note excellent preservation of epithelium, connective, and smooth muscle. In smooth muscle cells, thick filaments (120 Å in diameter) cut obliquely and organized in a quite regular fashion can be seen. × 11,800. (*Courtesy C. Peracchia*)

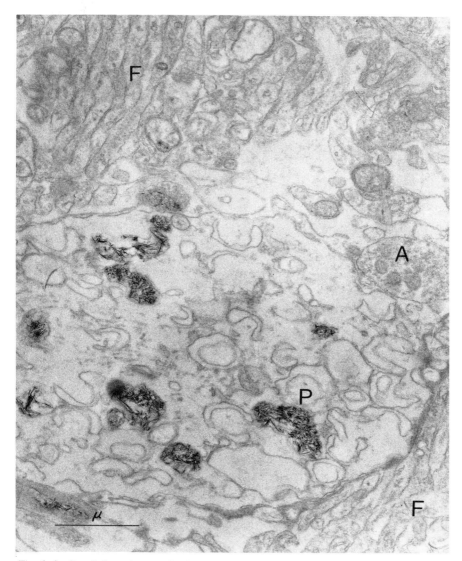

Fig. 1–2. Coupled succinate oxidation is demonstrated in rat cerebellar cortex. The tissue was incubated for 45 min prior to fixation. The mitochondria of the Purkinje-dendrite (P) contain heavy deposits of cupric ferrocyanide from the cytochemical reaction, whereas those of the contacting axon terminal (A) and of the parallel fibers (F) are without the reaction product. *(Courtesy F. Hajós)*

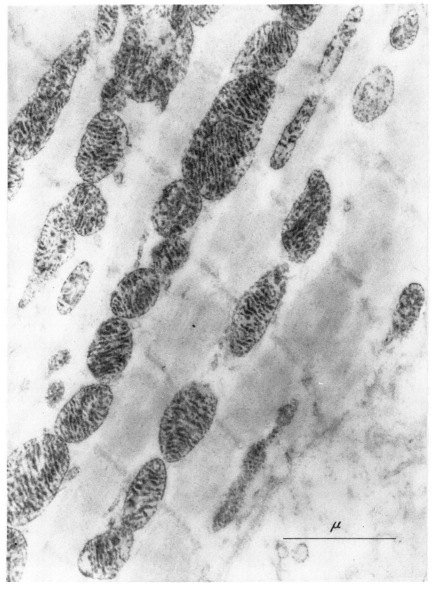

Fig. 1–3. Succinate dehydrogenase activity is demonstrated in the mitochondria of rat myocardium with the ferrocyanide method. The tissue was incubated for 60 min prior to fixation. (*Courtesy F. Hajós*)

Furthermore, certain enzymes (e.g., mitochondrial adenosine triphosphatase) remain inactive in unfixed sections and thus inaccessible to substrates; activation may be achieved by brief fixation with an aldehyde. It is also known that in unfixed ultrathin sections when the limiting membrane of certain organelles (e.g., lysosomes) is cut open during sectioning, the enzymes are released. A brief fixation prevents such leakage of enzymes. Finally, it should be noted that although cardiac muscle withstands incubation prior to fixation, the majority of the other tissue types (e.g., nervous tissue) fall apart, even with a short incubation prior to fixation.

Some enzymes in particulate specimens are readily inactivated, even by a brief fixation with low concentrations of aldehydes. Studies by Widnell (1972) showed that after a fixation of 15 min with 2% glutaraldehyde, 5'-nucleotidase was extensively inactivated ($\sim$ 95%) in plasma membranes and rough microsomes. However, the nonspecific phosphatase was inactivated to a much lower extent ($\sim$ 60%). In these studies, the enzymes were investigated in unfixed cell fractions. Apparently, the criterion for the preservation of fine structure is less stringent in this case than in intact tissue blocks. The effect of the duration of fixation of liver homogenate with glutaraldehyde on glucose-6-phosphatase activity is given in Fig. 1-4. This figure shows that the maximum enzymatic activity

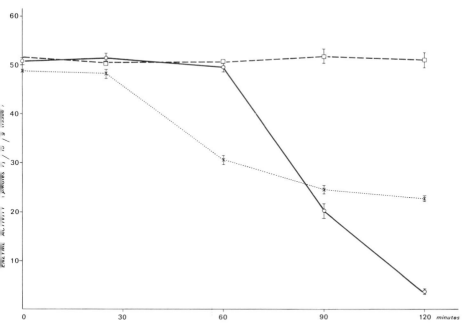

Fig. 1-4. The effect of the duration of contact between liver homogenate and glutaraldehyde on glucose-6-phosphatase activity before incubation. The dashed line represents the control values; the continuous line represents glutaraldehyde (15.10^{-2} μmoles/0.1 ml homogenate); the dotted line represents formaldehyde (3.10^{-1} μmoles/0.1 ml homogenate). *(Courtesy S. Casanova, M. Marchetti, C. Bovina and R. Laschi)*

is preserved when the duration of fixation does not exceed 30 min.

In conventional histochemistry, the adverse effects of incubation prior to fixation can be lessened somewhat by adding stabilizing agents to the incubation medium. Polyvinyl alcohol is a well-known membrane stabilizer, and has been proved effective in preventing the loss of nucleic acids and nitrogenous material (Altman *et al.,* 1966). Polyvinyl alcohol is especially effective in preserving the enzymatic activity related to mitochondria. Polypeptides derived from the partial degradation of collagen are more effective in demonstrating lysosomal activity. Stuart *et al.* (1969) were unable to show enzyme activity in whole blood cells incubated in the presence of polyvinyl alcohol, but could do so in the presence of a collagen polypeptide. Similarly, Butcher (1971) obtained maximum retention of glucose-6-phosphate dehydrogenase activity in the unfixed tissue sections in the presence of 30 to 40% polypeptide. The usefulness of these agents in electron cytochemistry should be tested.

A new approach to minimizing the diffusion of enzymes involves a semipermeable membrane which is interposed between the incubation solution and the tissue block. In this way, the diffusion of enzymes having a relatively high molecular weight is minimized, whereas substrate molecules with a relatively low molecular weight may diffuse through the semipermeable membrane. This procedure does not, however, prevent the diffusion of enzymes out of cell organelles.

The semipermeable membrane is prepared by soaking Visking dialysis tubing in a solution containing 0.1 gm EDTA in 250 ml of distilled water for 30 min. Following incubation, tissue specimens are fixed with aldehyde vapors at $20°C$ for $\sim$ 5 min. This method has been employed for the localization of lactate dehydrogenase (McMillan and Wittum, 1971) and acid phosphatases (Meijer, 1972) at the light microscope level. It remains to be seen if this method or a modified one can be useful in electron cytochemistry.

A relatively new approach to minimizing the loss of enzymes and diffusion artifacts caused by dehydration and infiltration with resins is frozen ultrathin sectioning (ultracryotomy). This method is simpler and faster than freeze-drying and freeze-substitution. Freezing renders the tissue sufficiently rigid so that it can be mounted in a microtome without the use of embedding resins. Ultracryotomy is difficult only with respect to actual sectioning. The basic principles governing ultracryotomy are now clearly established, and its possibilities and limitations are well understood.

It is possible to obtain frozen ultrathin sections of unfixed, briefly fixed, or conventionally fixed tissues. Hydrolytic enzymes have been demonstrated in ultrathin frozen sections of rat tissues briefly fixed with glutaraldehyde (Leduc *et al.,* 1967). Methods are also available for obtaining sections of the tissue which has not come in contact with any fixative, solvent, or flotation fluid (Christensen, 1971). This approach should allow enzymes to remain essentially in place and unaltered. Absence of fixation may, however, have undesirable repercussions, which have been discussed earlier.

As stated above, it is possible to employ simple, reproducible techniques for sectioning a wide variety of fixed or unfixed frozen specimens. It should be noted, however, that the sectioning characteristics of various types of frozen cells and tissue differ considerably compared with those of the same cells and tissues prepared by conventional methods. Thus, the optimal combination of parameters such as knife and clearance angles, cutting speed, and temperature is imperative for obtaining satisfactory sections of various cell and tissue types. Optimal conditions for electron cytochemistry using ultrathin frozen sections represent a compromise between many conflicting requirements unique to this technique.

Rate of Penetration

The rate of glutaraldehyde penetration is slower than that of formaldehyde; the K values for glutaraldehyde and formaldehyde are 0.34 and 2.0, respectively. The rate of penetration is dependent primarily upon the type of the tissue specimen and ambient temperature. Even similar tissue types obtained from different sources react differently to glutaraldehyde diffusion. Chambers *et al.* (1968) found that 4% glutaraldehyde penetrated rabbit liver and human liver 1.5 mm and 3 mm, respectively, in 9 hr at room temperature.

Penetration at room temperature is definitely faster than in the cold. The maximum penetration of 4% glutaraldehyde into human liver at room temperature and in the cold was reported to be 4.5 mm and 2.5 mm, respectively, in 24 hr (Chambers *et al.,* 1968). Therefore, in order to obtain uniform fixation of all cells in a tissue specimen within a period of two hours, the thickness of the specimen should be less than 0.5 mm. Prolonged fixation would not necessarily improve the quality of fixation, because penetration and fixation should be completed prior to the onset of autolytic changes.

The slow penetration rate of glutaraldehyde becomes a serious problem in the fixation of dense materials. This problem can be solved by using glutaraldehyde in combination with rapid penetrants such as paraformaldehyde. A mixture of glutaraldehyde (2%) and paraformaldehyde (2%) has been employed, for example, to preserve grooves (cleftlike invaginations) of the surface of plasmalemma of yeast cells (Ghosh, 1971). These grooves are not preserved when the cells have been fixed with glutaraldehyde alone. The rate of penetration can also be increased by adding dimethyl sulfoxide (DMSO) to the fixative, although its effects, as discussed below, on enzymes and ultrastructure are not fully known.

The depth of fixation is considered to be the same as the depth of penetration. Glutaraldehyde penetration into the tissue is indicated by a pale yellow color (which is primarily due to the formation of Schiff-positive bases when the dialdehyde reacts with basic amino acids) and by a firm texture. Since glutaraldehyde introduces Schiff positivity to tissues fixed in it, the presence of aldehyde groups in the region showing pale yellow color can be confirmed by its pink staining with Schiff's reagent.

Role of Dimethyl Sulfoxide

Dimethyl sulfoxide is a relatively nontoxic nonelectrolyte. Lovelock and Bishop (1959) introduced DMSO as a cryoprotective solute, and since then it has been used widely as a cryoprotectant. Dimethyl sulfoxide is considered to be an intracellular protective agent, in contrast to polyvinylpyrrolidone and dextran, which are nonpenetrating cryoprotectants.

The majority of the nucleated plant and animal cells are irreversibly damaged by the process of freezing and thawing. However, the damage to the fine structure can be minimized by treating the cells, prior to freezing, with solutions of cryoprotective nonelectrolytes such as DMSO or glycerol. Extensive extraction of nonmembranous constituents from unprotected hamster tissue-culture cells due to freezing and thawing can be minimized by treating these cells with DMSO (Bank and Mazur, 1972). Comparative ultrastructural studies of *in vitro* and *in situ* mitochondria indicate that DMSO reduces cryoinjury (Sherman, 1972).

According to Cope (1968), glycerol and ethylene glycol are superior to DMSO in preserving spatial relationships within the pancreatic tissue. He, therefore, recommended mixtures of DMSO and glycerol or DMSO and ethylene glycol, rather than DMSO alone.

It should be noted that DMSO does not prevent all morphological changes caused by freezing and thawing. Moreover, not all cells and tissues can withstand freezing in the presence or absence of DMSO. This is examplified by human granulocytes, which cannot withstand freezing in the presence or absence of DMSO, and the primary site of cryoinjury is the nucleus and not cytoplasm (Malinin, 1972). It must be emphasized that absence of ultrastructural alterations following freezing and thawing does not necessarily mean absence of injury at the subcellular level, since the resolving power of an electron microscope is limited. In this connection, it is important to remember that detectable ultrastructural alterations do not necessarily result in functional changes. It is well known that fragmented mitochondria retain important aspects of oxidative phosphorylation.

Cryoprotection by DMSO is partly due to its remarkable capacity for lowering the freezing point of aqueous solutions. When high concentrations of DMSO are present, there is less ice formation and a lower concentration of salts in the residual medium at temperatures below zero. The advantage of these properties becomes apparent when one considers that the main causes of damage to living cells during freezing and thawing are raised salt concentrations and ice formation. The colligative action of DMSO and other nonelectrolytes in reducing the concentration of salt in equilibrium with ice at any temperature is believed to be the primary mechanism by which they protect living cells at low temperatures.

That cryoprotection by DMSO may also be due to its buffering action has been proposed (Elford and Walter, 1972). When present extracellularly, it may prevent the denaturation of the cell membrane by acting as a buffer against the high concentration of damaging electrolytes (e.g., sodium chloride) in the bathing medium. In addition, DMSO may act as an intracellular pH buffer, and thus may help to prevent denaturation of intracellular proteins.

That DMSO postpones both the cell shrinkage and the onset of cation leaks to higher osmolarities has been demonstrated (Farrant, 1972). In this case, the onset of sodium or potassium leaks may be linked to cell shrinkage. It has been suggested that the efficacy of DMSO may be due to its ability to postpone cation leakage to higher osmolarities.

There is evidence to indicate that better preservation and staining of certain enzymes are obtained when specimens are treated with DMSO. This is accomplished by increasing the rate of penetration by the incubation media by adding DMSO. The majority of the incubation media penetrate the tissue at a rather slow rate; various components of the incubation medium probably penetrate at different rates. Copper ferrocyanide medium for succinic dehydrogenase, for instance, does not penetrate more than 20 μ in 30 min at room temperature. Moreover, the extent of reaction between the medium and the enzyme varies, depending upon the distance from the surface of the tissue block.

The prolonged incubation necessary for adequate penetration and interaction may result in the formation of artifacts and damage to tissue fine structure. The incubation time can be shortened by accelerating the rate of penetration of the incubation medium. This can be accomplished, as stated above, by adding DMSO to the incubation medium. Incubations at temperatures lower than usual and the addition of sucrose (Holt and Withers, 1958), polyvinylpyrrolidone (Novikoff, 1956), and glycerol (Melnick, 1967) (as high as 15%) have also been attempted in the hope of improving the preservation of cell structure.

Enhanced staining of enzymes has been demonstrated by adding DMSO to the incubation medium (Reiss, 1971). Less heterogeneous activity of succinic dehydrogenase has been obtained in the mitochondria of cardiac and skeletal muscles of the mouse by adding 3 to 5% DMSO to the copper ferrocyanide incubation medium (Makita and Sandborn, 1971). Dimethyl sulfoxide was added to napththol AS BI phosphate medium for the localization of acid phosphatase activity in rat and slug liver (Bowen, 1971). The activity of these enzymes is apparently not inhibited by DMSO. Studies of the effect of DMSO on the incubation media for other enzymes are in order. It should be noted that the addition of DMSO results in higher osmolarity of the incubation medium.

An alternative approach is to treat the tissue specimens with DMSO prior to incubation. Such an approach has been utilized by Göthlin and Ericsson (1971) to obtain localization of acid phosphomonoesterase activity in the brush border osteoclasts. They immersed the fracture callus into 20% DMSO for 24 to 48 hr prior to incubation. Another approach is to fix the tissue in the presence of DMSO. The addition of DMSO to the fixative has been shown to result in improved preservation of frozen tissues (Etherton and Botham, 1970).

Several explanations have been offered for the improved preservation of enzymatic activity due to the addition of DMSO. That DMSO increases the permeability of cell and/or organelle membranes has been indicated (Chang and Simon, 1968; Ghajar and Harmon, 1968; Misch and Misch, 1968; Gander and Moppert, 1969). According to Hanker et al. (1970), DMSO facilitates electron transfer from the thiol groups of dehydrogenases to tetrazolium acceptors.

Overwhelming evidence indicates that DMSO penetrates cells rapidly, and thus exerts its effect upon cellular sites in addition to cell membranes (Bank and Mazur, 1972), although it has been suggested that DMSO is restricted to the surface of cell membranes and intercellular spaces, and thus its role in enhancing penetration of the incubation medium may not be an intracellular phenomenon. It is admitted that very little is known regarding the mechanisms responsible for the stabilizing action of DMSO on enzymes.

A brief comment on the use of DMSO in conventional histochemistry is in order. Dimethyl sulfoxide has been used extensively for preserving tissue structure and enzymatic activity in conventional histochemistry. Recent studies of the effects of DMSO on histochemical reactions indicate that cell morphology and enzymatic activity are better preserved during freezing in the presence of DMSO (Ward and Smith, 1971). Living cells and tissues of many kinds have been "banked" at very low temperatures in media containing DMSO (Pegg, 1970). After thawing and removal of DMSO, a high proportion of the cells resume activity. Bank and Mazur (1972) reported that hamster tissue culture cells rapidly frozen and thawed in the presence of 0.4% DMSO showed ~ 80% survival rate.

Caution is warranted in the use of DMSO in electron microscopy, for the production of artifacts by this reagent cannot be overlooked. It has been demonstrated, for instance, that the action of DMSO at the cell surface may produce blebbing in hepatocytes (Shilkin et al., 1971). It has also been shown that even a brief (2 min) perfusion of isolated rat heart with DMSO resulted in ultrastructural alterations; the effect was immediate on the T-system and mitochondria (Feuvray and De Leiris, 1973). The degree of alterations was dependent upon DMSO concentration. These alterations may be due to the hypertonic effect of DMSO. However, in other tissues such as rat cerebral cortex, DMSO treatment does not alter the fine structure (Brunk and Ericsson, 1972a).

Temperature

Generally higher temperatures enhance the speed of chemical reactions between the fixative and the enzyme. Higher temperatures increase the rate of penetration (diffusion) of the fixative into the tissue and the rate of autolytic changes simultaneously. Higher temperatures also increase the rate of penetration of the incubation medium into the tissue specimen and the speed of its reaction with the enzyme. Generally, fixation for electron cytochemistry of enzymes is carried out at 4°C.

It is known that enzyme activation is much lower at 4°C than at room temperature. However, vascular perfusion at 4°C may result in poor preservation of the fine structure, partly because of vascular spasm. This problem can be overcome by perfusing the tissue with 0.1% solution of Procain in a buffer for 3 min at 20°C prior to infusion with the cold fixative. An osmolality of 300 mOsM of this solution is recommended for central nervous tissue. This technique has been described in detail by Forssmann *et al.* (1967).

Concentration of Aldehydes

Although the optimal concentration of an aldehyde to obtain maximum preservation of a given enzymatic activity in a given tissue is determined only by experimentation, a general range of concentration applicable to the majority of the enzymes has been established; a concentration range of 0.5 to 6% is considered satisfactory. According to Nagata and Murata (1972), the preservation of lipase activity in the pancreatic acinar cells of mice and rats was not affected when the specimens were fixed with 2.5 to 5.0% glutaraldehyde.

It is known that fixation with low concentrations (0.2 to 0.5%) of glutaraldehyde strengthens the binding of certain enzymes to membranes. Studies by Ellar *et al.* (1970), for instance, indicate that after fixation with 0.5% glutaraldehyde, the release of enzymes such as ATPase, NADH dehydrogenase, and polynucleotide phosphorylase from the plasma membrane of *Micrococcus lysodeikticus* by washing with buffers is prevented, and that these enzymes become more strongly attached to the membranes. Furthermore, the cross-linking action of glutaraldehyde does not affect the activity of these enzymes.

In general, higher concentrations of aldehyde solutions destroy enzymatic activity and damage the cellular fine structure. The maximum enzymatic activity is recovered after treatment with low concentrations of aldehydes for moderate periods of time. These conditions permit sufficient diffusion of the aldehyde into the tissue specimen but limit the extent of cross-linking. High concentrations of aldehydes and long durations of fixation result in decreased enzymatic activity due to excessive cross-linking.

Various cell organelles differ in their response to variations in the concentration of an aldehyde fixative. Baker and McCrae (1966) indicated that low concentrations (0.25%) of formaldehyde disrupted the endoplasmic reticulum, whereas mitochondria were insensitive to changes in concentration. However, mitochondria are generally more sensitive to changes in the osmolality of the solutions.

The pH

The pH of the fixative solution is a critical factor in the maximal preservation of enzymatic activity. Not only do different cells and tissues require a specific pH, but various organelles in the same cell also react differently at a given pH with regard to the preservation of enzymatic activity. In other words, the preservation of an enzyme in an organelle may be maximum at a given pH, but other organelles may not show maximal activity at the same pH. It has been shown that the optimal pH for demonstrating glucose-6-phosphatse in the endoplasmic reticulum, nuclear envelope, and dictyosomes in cultured chicken heart cells range between 5.8 and 6.2, whereas in the nucleus the reaction product is noticed only at pH 6.8 (Schäfer and Hündgen, 1971). Similar information on other enzymes and tissues is needed. Small changes in the pH of a buffer or a fixative may bring about large changes inside the cell, including the activity of enzymes.

Buffers

The role played by buffers in the preservation of enzymatic activity has not been fully appreciated. It is quite likely that a buffer wash prior to and/or following fixation with aldehydes alters the enzymatic activity. Phosphate buffer is known to completely inhibit the activity of glucose-6-phosphate dehydrogenase (Löhr and Walker, 1963), while cacodylate buffer tends to inhibit β-glucuronidase activity somewhat (Smith and Fishman, 1968). Lactate dehydrogenase shows $\sim$ 25% lower activity in 0.4 M Tris-HCl buffer (pH 7.4) than in 0.1 M phosphate buffer (pH 7.4) (Dahl and From, 1971). This reduced activity in the former buffer may be due to conformational changes in the enzyme molecule.

It is appropriate to mention that although information on enzyme activities obtained through biochemical studies is significant, enzymes in an isolated state may possess different sensitivity to reagents, including buffers, from that *in situ*. It has been pointed out that a 35% difference in succinate dehydrogenase activity measured with biochemical methods is necessary to yield an observable difference in the tissue sections at the light microscope level (Riecken *et al.*, 1969). No such information is available at the subcellular level.

Little information is available regarding the effects of a buffer wash of variable durations on the activity of various enzymes. Such information is apparently important in understanding the role played by buffers in the fixation for enzyme localization. Furthermore, not only the type of ions in the buffer but also the electrolytes added to the fixation solutions exert a profound effect upon enzyme activities. It cannot be assumed that various soluble enzyme systems in cells respond the same way to different buffer systems.

Arborgh *et al.* (1971) have reported that immersion of the tissue following prolonged fixation with glutaraldehyde in a chilled buffer (0.1 M Tris-maleate) resulted in a significant increase in the activity of certain enzymes (e.g., acid phosphatase and aryl sulfatase). A buffer wash, however, may inhibit the activity of other enzyme systems.

A prolonged wash in buffer between the prefixation with an aldehyde and incubation and postfixation with osmium tetroxide tends to produce undesirable effects on the fine structure. These effects include an uneven fixation, cell shrinkage with widened extracellular spaces, extraction or swelling of mitochondrial matrices, and clumping of chromatin (Trump and Ericsson, 1965; Trump and Bulger, 1966). It has also been shown that the storage of glutaraldehyde-perfused tissue for as little as ½ hr in cacodylate buffer has a profound flattening effect on agranular synaptic vesicles of cholingergic nerve endings (Bodian, 1970). The effects of various buffers on the fine structure have been discussed in detail (Hayat, 1970).

Osmolality

The direct effect of the fixative osmolality on the appearance of fine structure has been well established. The osmolality of the fixative solution and the presence of salts in the incubation medium seem to affect the cytochemical reaction for the demonstration of enzymatic activity. It has been demonstrated that fixation at an improper osmotic pressure resulted in enzyme diffusion from the lysosomes in *in vitro* cultured cells (Brunk and Ericsson, 1972b). At least one of the hydrolytic enzymes, acid phosphatase, can disappear from morphologically intact lysosomes, probably as a result of hypo-osmotic damage (Brunk and Ericsson, 1972c). Thus, enzymes can escape to the cell sap through uninterrupted membranes.

The presence or absence and the concentration of electrolytes or nonelectrolytes in various solutions may affect the speed of reaction and substrate specificity. Recent studies of ATPase indicated that the presence of high concentrations of chloride salts in the incubation medium resulted in a rapid histochemical reaction and that the enzymatic activity showed up at some additional sites (Berg *et al.*, 1972). A rapid localization of ATPase activity was obtained with incubation mixtures containing 2 M chlorides of monovalent cations. It is thought that the addition of chloride salts decreases the concentration of lead required to saturate the reaction mixture, and this effect accounts for the acceleration in the cytochemical reaction.

Malic enzyme is another system the stabilization of which has been demonstrated to be affected by monovalent cations. It was shown that this enzyme from *Halobacterium cutirubrum* required monovalent cations for both activation and stabilization (Cazzulo and Vidal, 1972). It has been suggested that the role of salts in enzyme activation may differ from the role of salts in enzyme stabilization (Hubbard and Miller, 1969). It must be pointed out that salt concentrations higher than optimal will inhibit enzymatic activity.

Sucrose and polyvinylpyrrolidone are also known to affect enzymatic activity. Even very low concentrations of sucrose or polyvinylpyrrolidone tend to inhibit the activity of some cerebral enzymes (e.g., mitochondrial succinate dehydrogenase) (Lisý et al., 1971). This inhibition does not seem to be related to the increased osmolality caused by the addition of nonelectrolytes, because even isotonic solutions have an inhibitory effect. The exact reasons for this phenomenon are not known. Hinton et al. (1969) have discussed certain theoretical reasons for the inhibitory effect of sucrose on enzyme activity.

It is pointed out that nonelectrolytes are more than osmotic agents to which the cell is impermeable, because their presence greatly increases the effect of the salts subsequently added. The possibility of sucrose reaction by hydrogen bonding with groups on the membrane has been proposed (Bernheim, 1971). From the foregoing discussion it is apparent that a number of varied and complex factors are involved in the total effects of osmotic protective agents on the preservation of enzymatic activity and ultrastructure. Hayat (1970) has discussed in detail the possible mechanisms involved in the effects of various electrolytes and nonelectrolytes on the preservation of cellular components.

Average osmolalities of phosphate-buffered glutaraldehyde formulations of different concentrations are given below. The commercially available glutaraldehyde was purified by charcoal treatment, and the osmolalities were measured with a commercial osmometer (Chambers et al., 1968).

	Milliosmoles	pH
Phosphate buffer (0.1 M)	230	7.4
1.2% glutaraldehyde	370	7.2–7.3
2.3% glutaraldehyde	490	7.2–7.3
4% glutaraldehyde	685	7.1–7.3

Final osmolalities of given concentrations of various aldehydes containing sucrose are given in Table 1-4.

Autolytic Changes

The maximal preservation of enzymatic activity is apparently obtained when the tissue is processed in situ in the living organism. If this is not practical, the tissue should be processed immediately following its removal from the organism. However, in certain instances such as biopsy, surgical, and postmortem specimens, a considerable amount of time may elapse between specimen collection and processing. Studies on the postmortem stability of enzymes indicate that the falloff in enzyme activity is not as rapid as has been thought previously. Tyrer et al. (1971) have demonstrated that in rabbit brain and spinal cord, succinate dehydrogenase activity remained stable after 24 hr postmortem exposure at 37°C, while lactate dehydrogenase, NAD diaphorase, and monoamine oxidase

Table 1-4 Final Osmolalities of Given Concentrations
of Various Aldehydes and Sucrose Solutions

	Aldehyde concentration	*Sucrose concentration*	*Final osmolality*
Glutaraldehyde	0.20 M =2.00%	0.20 M =6.9%	480 mOsM
Acrolein	0.20 M =1.12%	0.45 M =15.4%	720 mOsM
Hydroxyadipaldehyde	0.46 M =6.00%	0.20 M =6.9%	760 mOsM
Paraformaldehyde	1.33 M =4.00%	0.12 M =4.0%	1280 mOsM
Methacrolein	0.20 M =1.40%	0.29 M =10.0%	520 mOsM
Crotonaldehyde	0.55 M =3.90%	0.45 M =15.4%	1130 mOsM
Glyoxal	0.20 M = 1.16%	0.35 M = 12.0%	620 mOsM
Acetaldehyde	0.45 M =1.96%	0.53 M =18.0%	800 mOsM

From N. Weissenfels *et al.* (1971). Used by permission.

activities were less stable at $37°C$ but were stable at $22°C$. Glutamate and glucose-6-phosphate dehydrogenase activities fell significantly with storage of tissue at $22°C$.

The above-mentioned and other data suggest that valid results can be obtained by studying tissue specimens several hours following the death of the organism. It also seems possible to carry out valid cytochemical studies on certain enzymes in human postmorten specimens. As expected, the falloff in enzyme activity increases with increasing temperature of postmortem exposure prior to freezing. The postmortem stability of cell organelles has been discussed in detail elsewhere (Hayat, 1970).

EFFECTS OF DIETARY AND OTHER FACTORS

Evidence is available which indicates that the activity of some enzymes in laboratory animals undergoes changes due to variations in food intake. Moreover, changes in the activity of isoenzymes may differ in different tissues of an animal. It is known that hormonal function can be affected by the composition of the food, and that changes in hormonal function, in turn, play a significant role in the induction of some enzymes. In fact, the dependence of enzymatic activities upon the nutritional state of an organism has been used for elucidating the physiological role played by enzymes in metabolic pathways.

It has been reported that rat liver ornithine decarboxylase undergoes a profound diurnal change in activity due to cyclic variation in the amount of protein uptake (Hayashi *et al.*, 1972). Studies of acetyl-CoA synthetase activities in different organs of normal, fasted, and carbohydrate-refed rats indicate that in liver cytoplasm and epididymal fat tissue the activity of this enzyme decreased to ~ 50% of normal after fasting (Barth *et al.*, 1972). Refeeding a carbohydrate-rich diet caused a rise of as much as 170% of the fasting activity in liver and a rise of 550% in the fat tissue. That NADP-linked malic enzyme and glycero-kinase show diurnal rhythms in the liver of cockerels has been shown (Chandra-bose *et al.*, 1971). Attempts have been made to correlate diurnal variations in enzyme activities with corresponding rhythmicities in hormones known to be inducers of these enzymes (Civen *et al.*, 1967).

Quantitatively significant seasonal variations in the activities of enzymes in human beings are well known. That erythrocyte glucose-6-phosphate dehydrogenase is subject to seasonal variations in healthy persons has been demonstrated (Hilgertová *et al.*, 1972). A steep rise in this enzyme was observed in the winter of 1969-70, which was excessively cold and long in Prague. This rise in the enzyme is thought to be due to enhanced thyroid activity as a result of adaptation to cold.

It should be noted that many experimental treatments to which tissues and animals are exposed have a significant effect on the enzymatic activity. Hepatectomy or the administration of growth hormone, for instance, influences enzymatic activity. It is apparent from the above discussion that cognizance of the physiological state of the animal is warranted when interpreting activities of inducible enzymes in tissues.

COMMERCIAL GLUTARALDEHYDE

Commercial glutaraldehyde is unsuitable for the localization of activity of the majority of the enzymes because it rapidly inactivates enzymes during fixation. This undesirable property is, in part, due to the presence of impurities such as glutaric acid, acrolein, glutaraldoxime, ethanol, and various polymers and products of oxidation and photochemical degradation. It should be noted that the presence of impurities also influences the pH, osmolality, and concentration of glutaraldehyde.

Recent evidence obtained from nuclear magnetic resonance spectroscopic and chromatographic studies indicates that commercial glutaraldehyde is largely polymeric, and contains dimer, cyclic dimer, trimer, bicyclic trimer, and other polymeric species. The principal polymer in commercial glutaraldehyde is considered to be α-β-unsaturated dimer, which forms slowly after ordinary storage conditions and absorbs in the ultraviolet at 235 nm. The polymerization of this dimer is enhanced by heat, acids, and bases. This is the dimer which is partly responsible for the rapid inactivation of enzymes during fixation, because it readily cross-links with enzyme proteins. For this reason, purified glutaraldehyde is recommended for enzyme localization.

There is evidence which indicates that the presence of a high concentration of inorganic phosphate in commercial glutaraldehyde may confuse the localization of phosphatase activity. One way to remove this phosphate is to subject the fixed specimens to dialysis. Sexton *et al.* (1971) successfully removed the phosphate from homogenates of tissue fixed with glutaraldehyde by dialysis in 30 cm of 1 cm dialysis tubing against 2 lit of distilled water for 2 hr at $4°C$; no loss in β-glycerophosphatase activity was detected after this dialysis procedure.

PURIFICATION OF GLUTARALDEHYDE

Using various methods, several workers (Clift and Cook, 1932; Fein and Harris, 1962; Fahimi and Drochmans, 1965; Smith and Farquhar, 1966; Anderson, 1967; Hopwood, 1967b; Frigerio and Shaw, 1969) have carried out purification and determination of concentration of glutaraldehyde with varying degrees of success. In the technique developed by Fahimi and Drochmans (1965), commercial glutaraldehyde is treated with activated charcoal followed by vacuum distillation. The distillate thus obtained is a clear, oily solution of a high degree of purity and concentration, which can be diluted to the desired concentration with boiling distilled water. The concentration of the distillate is determined with a recording spectrophotometer. Glutaraldehyde shows absorption maximum in the ultraviolet at wavelength of 280 nm; absorption at any other wavelength is caused by impurities. Since a direct linear relationship is found between the osmolality and the optical density at 280 nm, this density can serve to determine both the concentration and osmolality of the distilled glutaraldehyde (Fahimi and Drochmans, 1965).

Hopwood (1967b) indicated that glutaraldehyde could be fractionated from the impurities on Sephadex G-10. Anderson (1967) analyzed glutaraldehyde by hydroxylamine titration, and indicated that a single-stage vacuum distillation under moderate vacuum yields glutaraldehyde of equivalent purity. This method is rather lengthy and thus difficult to use for a day-to-day control of fixing solutions. It should be noted that repeated washings with alkaline charcoal result in the elevation of pH of the glutaraldehyde solution. Recently, Frigerio and Shaw (1969) developed a method specific for the determination of aldehyde group in glutaraldehyde. This method is simple and rapid, and can be easily applied to determine the rate of glutaraldehyde deterioration under fixation conditions. In this method, the concentration of glutaraldehyde in aqueous solution or in fixing solutions is determined by measuring the amount of unreacted bisulfite.

STORAGE OF GLUTARALDEHYDE

In order to prevent reoxidation, purified glutaraldehyde should be stored under nitrogen at low temperature. In fact, purified glutaraldehyde remains relatively stable for several months if stored under atmospheric conditions at $4°C$, providing that the pH is lowered to about 5.0. The desired pH for storing can be obtained by adding a few drops of HCl. Unpurified glutaraldehyde should be stored at neutral pH as a dilute solution (4% or less) at subfreezing temperatures.

FIXATION

Fixation is carried out either with glutaraldehyde, with paraformaldehyde, or with a mixture of these two aldehydes. Both immersion and vascular perfusion have been employed for fixation. Fresh tissue blocks (less than 1 mm^3) can be directly fixed with an aldehyde solution. The prefixed blocks are cut freehand into $\sim$ 0.2 mm^3 blocks or are cut at 40 to 80 μ on a Smith-Farquhar microchopper, on OmE Reichert microtome fitted with the Frigistor system, or on Oxford Vibratome; the latter procedure permits a more satisfactory penetration of substrates and reagents into the tissue. Alternatively, tissue blocks can be embedded in 2% agar prior to cutting on the microchopper. These thin tissue sections are again fixed in a cold aldehyde solution, and then washed thoroughly with an appropriate buffer. The exact details of fixation for each enzyme are given in subsequent chapters.

Tissue sections or blocks are incubated in an appropriate medium, the details of which for each enzyme can be found in the first and second volumes of this work. The incubated specimens are thoroughly washed with a buffer and postfixed with buffered-osmium tetroxide (1 to 2%) for ½ to 1 hr at 4°C. In some procedures, potassium permanganate (1 to 2%) can be employed as a postfixative instead of osmium tetroxide. Young embryos of rabbits, for instance, have been postfixed with potassium permanganate for 30 min at 4°C following incubation for the localization of acetylcholinesterase activity (Tennyson et al., 1971).

In using an alternative procedure, fresh tissue blocks are frozen rapidly in isopentane cooled to -70°C with dry ice and cut in a cryostat at -13° to -10°C to obtain sections (20 to 40 μ). These sections are then thawed at room temperature and immersed immediately in an aldehyde solution for fixation. This procedure invariably produces some damage to tissue fine structure by the freezing and thawing cycle.

INCUBATION

The details regarding the composition and application of incubation media are given for each enzyme in subsequent chapters. A brief general comment on incubation parameters will suffice here. The exact composition of the incubation medium and the conditions for its application are of utmost importance in obtaining accurate localization of enzymatic activity. An exact knowledge of the ingredients, pH, temperature, osmolality, and duration of application is necessary. Overlapping substrate affinities are not uncommon, and even a slight departure from optimal cytochemical conditions may confuse the proper identification of a specific enzyme activity.

That the concentration of various ions in the incubation media influences the stabilization, activation, or inactivation of enzymatic activities has been known for a long time. In this connection, although biochemical studies have provided the bulk of the evidence, the effects of ions on tissue blocks have also been well documented. Recent studies by Dantzler (1972), for instance, indicated that a prolonged (2 hr) incubation of chicken kidney slices in potassium-free medium produced significant residual depression in Na-K-ATPase activity in heavy microsomes. Similar incubation of snake kidney, however, had no residual depression on the enzyme activity. Incubation of chicken and snake kidney slices in low-sodium medium showed no residual depression effect on the activity of this enzyme.

The introduction of enzyme activators and inhibitors such as calcium, magnesium, and sodium fluoride may alter the critical concentration of the capture agents, which will, in turn, cause changes in the pattern of localization unrelated to changes in enzyme activity (Moses and Rosenthal, 1968). It is apparent that not only the pH but also the type of ions present in the buffer, fixative solutions, and incubation medium exert an influence on the demonstrable enzyme activity. For a comprehensive discussion on the role of cations in the preservation of cell fine structure, the reader is referred to Hayat (1970).

In order to obtain maximum reaction between the enzyme and the substrate, it is imperative to establish the optimal duration of incubation. Both shorter and longer durations of incubation will result in a suboptimal localization of the reaction product. The ideal duration can be determined by monitoring nonfrozen sections with a light microscope during the course of incubation.

ARTIFACTS

The possibility of diffusion and adsorption of soluble enzyme protein during preparatory procedures, especially during fixation, is now universally accepted. Moreover, it is not always easy to decide, for instance, whether cytochemical staining is due to phosphatase activity acting on endogenous substrate or to adsorption of lead. Noncrystalline appearance of the staining may be one of the indications of nonparticipation of an endogenous substrate. In the case of ATPase activity, lead used as a trapping agent not only inhibits the enzyme but also hydrolyzes the substrate. Moses and Rosenthal (1968) have discussed the pitfalls in the use of lead ion for the localization of certain ATPases. Artifacts may be produced by lead adsorption, by adsorption of cytoplasmic enzymes to organelles during fixation, and by inactivation or activation of enzymes by glutaraldehyde.

Uneven penetration of the incubation medium is probably the most frequent cause of inconsistent staining. The speed of penetration is, in part, affected by the tonicity of the medium with reference to the osmotic pressure within the tissue. It should be remembered that fixation with aldehydes does not completely destroy the selective permeability of membranes. Difficulties with penetration, however, can be minimized by incubating thin sections (125 to 250 μ) of aldehyde-fixed tissues. Gahan *et al.* (1967) indicated that the problem of penetration can be overcome by using cryostat-cut frozen sections. The role of DMSO in facilitating the penetration of the media has been discussed earlier. The production of artifacts during tissue processing has been discussed by Deane (1963), De Jong *et al.* (1967), Gahan (1965), Hayat (1970), Essner (in this volume), and Iversen (in this volume).

Correct interpretation of cytochemical localization of enzymes is dependent upon the use of adequate controls. The importance of adequate positive and negative controls in enzyme cytochemistry cannot be overemphasized. Careful biochemical determinations, run in parallel for each specific enzyme, can be utilized as a double check.

EMBEDDING

Prior to embedding in water-immiscible resins (e.g., Epon, Araldite, Vestopal, and Spurr mixture), tissue specimens are dehydrated rapidly in a series of increasing concentrations of acetone or ethanol solutions. Dehydration with ethanol must be followed by one or two quick changes in 100% propylene oxide, because ethanol is immiscible with many resins. Acetone, on the other hand, is miscible with most resins. Dehydration is followed by a gradual infiltration by the resin mixture. The duration of infiltration is determined by the size and density of the tissue specimen and by the viscosity of the resin mixture. Two typical procedures for dehydration, infiltration, and polymerization are given below.

I.
1. 10% acetone — 4 min
2. 30% acetone — 4 min
3. 50% acetone — 4 min
4. 95% acetone — 4 min
5. 100% acetone (2 changes) — 4 min
6. 100% acetone and resin (1:1) — ½ hr
7. 100% acetone and resin (1:3) — ½ hr
8. resin — ½ hr
9. resin (polymerize at 60°C) — 24 hr

II.
1.	10% ethanol	4 min
2.	30% ethanol	4 min
3.	50% ethanol	4 min
4.	95% ethanol	4 min
5.	100% ethanol	4 min
6.	100% propylene oxide (2 changes)	4 min
7.	100% propylene oxide and resin (1:1)	½ hr
8.	100% propylene oxide and resin (1:3)	½ hr
9.	resin	½ hr
10.	resin (polymerize at 60°C)	24 hr

The advantage of using water-miscible resins (e.g., Durcupan and glycol methacrylate) is that tissue specimens are embedded without coming in contact with organic solvents. The technique involves the use of water-miscible resins as the dehydration agents and their gradual blending and replacement by water-immiscible resins. Alternatively, tissue specimens are cut while embedded in water-miscible resins. These resins, however, may themselves act as solvents of some lipids, proteins, and carbohydrates. A typical procedure for dehydration and infiltration is given below.

1.	50% glycol methacrylate	15-30 min
2.	70% glycol methacrylate	15-30 min
3.	90% glycol methacrylate	15-30 min
4.	100% glycol methacrylate	30-60 min
5.	100% glycol methacrylate and Epon (1:1)	30-60 min
6.	100% glycol methacrylate and Epon (1:3)	30-60 min
7.	Epon (2 changes)	1-2 hr
8.	Epon (polymerize at 60°C)	24 hr

For procedures using other water-miscible embedding media, the reader is referred to Hayat (1970 and 1972).

EMBEDDING MEDIA

Araldite

Araldite 502	68 ml
DDSA	19 ml
TAC (triallyl cyanurate)	10 ml
DMP-30	3 ml

A decrease in the proportion of DDSA will result in a softer block.

Epon

Mixture A:

Epon 812	5 ml
DDSA	8 ml

Mixture B:

Epon 812	8 ml
NMA	7 ml

Final Embedding Mixture:

Mixture A	13 ml
Mixture B	15 ml
DMP-30	16 drops

An increase in the proportion of Mixture B will result in a harder block.

Glycol Methacrylate

Mixture A:

Glycol methacrylate	97 ml
Distilled water	3 ml

Mixture B:

Butyl methacrylate	98 ml
Luperco	2 ml

Final Mixture:

Mixture A	7 parts
Mixture B	3 parts

Maraglas

Maraglas 655	48 ml
Cardolite NC-513	40 ml
TAC	10 ml
DMP-30	2 ml

A decrease in the proportion of TAC will result in a harder block.

Spurr Mixture

ERL 4206	10 gm
DER 736	6 gm

NSA (nonenyl succinic anhydride)   26   gm
S-1 (dimethylaminoethanol)   0.4 gm

An increase in the proportion of DER 736 will result in a softer block.

Vestopal

Vestopal   100 ml
Benzoyl peroxide   1   ml
Cobalt naphthenate   0.5 ml

Glutaraldehyde-Carbohydrazide (GACH) (Heckman and Barrnett, 1973)

This polymer has been introduced to avoid extraneous density by osmium tetroxide and lipid loss by organic solvents. The activity of some "hardy" enzymes may survive GACH embedding so that they may be localized in ultrathin sections. The polymer is prepared by adding carbohydrazide to 50% glutaraldehyde to yield a final concentration of 150 mg/ml. The former is added in three portions to the glutaraldehyde in a beaker, which is kept in an ice bath on a magnetic stirrer. The solution is stirred rapidly during each of the three additions and for 15 min or longer between additions. The required pH can be obtained by adding 1 N NaOH.

REFERENCES

Ahrens, R., and Weissenfels, N. (1969). Nachweis von Adenosintriphosphatase inden Mitochondrien Kultivierter Hühnerherzmyoblasten. *Histochemie* **19**, 248.

Aloyo, V. J., Geller, A. M., Marcus, C. J., and Byrne, W. L. (1972). Properties of glutaraldehyde-modified bovine hepatic fructose-1,6-diphosphatase. *Biochim. Biophys. Acta* **289**, 242.

Altman, F. P., Butcher, R. G., and Chayen, J. (1966). The retention of cell components and structure in unfixed sections during histochemical investigations. *Proc. Roy. Micros. Soc.* **1**, 127.

Anderson, P. J. (1967). Purification and quantitation of glutaraldehyde and its effects on several enzyme activities in skeletal muscle. *J. Histochem. Cytochem.* **15**, 652.

Arborgh, B., Ericsson, J. L. E., and Helminen, H. (1971). Inhibition of renal acid phosphatase and aryl sulfatase activity by glutaraldehyde fixation. *J. Histochem. Cytochem.* **19**, 449.

Avrameas, S., and Ternyck, T. (1969). The crosslinking of proteins with glutaraldehyde and its use for the preparation of immunoabsorbents. *Immunochem.* **6**, 53.

Baker, J. R., and McCrae, J. M. (1966). The fine structure resulting from fixation by formaldehyde: the effects of concentration, duration, and temperature. *J. Roy. Micros. Soc.* **85**, 391.

Bank, H., and Mazur, P. (1972). Relation between ultrastructure and viability of frozen-thawed Chinese hamster tissue-culture cells. *Expt. Cell Res.* **71**, 441.

Barrnett, R. J., Perney, D. P., and Hagström, P. E. (1964). Additional new aldehyde fixatives for histochemistry and electron microscopy. *J. Histochem. Cytochem.* **12**, 36.

Barrow, P. C., and Holt, S. J. (1971). Differences in distribution of esterase between cell fractions of rat liver homogenates prepared in various media. *Biochem. J.* **125**, 545.

Barth, C., Sladek, M., and Decker, K. (1972). Dietary changes of cytoplasmic acetyl-CoA synthetase in different rat tissues. *Biochim. Biophys. Acta* **260**, 1.

Berg, G. G., Lyon, D., and Campbell, M. (1972). Faster histochemical reaction for ATPase in the presence of chloride salts, with studies of the mechanism of precipitation. *J. Histochem. Cytochem.* **20**, 39.

Bernheim, F. (1971). The effects of alcohol and sugars on the swelling rate of cells of *Pseudomonas aeruginosa* in various salts. *Microbios* **4**, 49.

Blough, H. A. (1966). Selective inactivation of biological activity of Myxoviruses by glutaraldehyde. *J. Bact.* **92**, 266.

Bodian, D. (1970). An electron microscopic characterization of classes of synaptic vesicles by means of controlled aldehyde fixation. *J. Cell Biol.* **44**, 115.

Bowen, I. D. (1971). A high resolution technique for the fine structural localization of acid hydrolases. *J. Microscopy* **94**, 25.

Brunk, U. T., and Ericsson, J. L. E. (1972a). Electron microscopical studies on rat brain neurons. Localization of acid phosphatase and mode of formation of lipofuschin bodies. *J. Ultrastruct. Res.* **38**, 1.

Brunk, U. T., and Ericsson, J. L. E. (1972b). Demonstration of acid phosphatase in *in vitro* cultured cells. Significance of fixation, tonicity and permeability factors. *Histochem. J.* **4**, 349.

Brunk U. T., and Ericsson, J. L. E. (1972c). Cytochemical evidence for the leakage of acid phosphatase through ultrastructurally intact lysosomal membranes. *Histochem. J.* **4**, 479.

Butcher, R. G. (1971). Tissue stabilization during histochemical reactions: the use of collagen polypeptides. *Histochemie* **28**, 231.

Cartensen, E. L., Aldridge, W. G., Child, S. Z., Sullivan, P., and Brown, H. H. (1971). Stability of cells fixed with glutaraldehyde and acrolein. *J. Cell. Biol.* **50**, 529.

Casanova, S., Marchetti, M., Bovina, C., and Laschi, R. (1972). A study of the effects of fixation on liver glucose-6-phosphatase activity for electron microscope cytochemistry, *J. Submicro. Cytol.* **4**, 261.

Cazzulo, J. J., and Vidal, M. C. (1972). Effect of monovalent cations on the malic enzyme from the extreme halophile, *Halobacterium cutirubrum. J. Bact.* **109**, 437.

Chambers, R. W., Bowling, M. C., and Grimley, P. M. (1968). Glutaraldehyde fixation in routine histopathology. *Arch. Path.* **85**, 18.

Chandrabose, K. A., Bensadoun, A., and Clifford, C. K. (1971). Diurnal rhythms of liver enzymes in the chicken, *Gallus domesticus. Inter. J. Biochem.* **2**, 581.

Chang, C.-Y., and Simon, E. (1968). The effects of dimethyl sulfoxide (DMSO) on cellular systems. *Proc. Soc. Exp. Biol.* **128**, 60.

Christensen, A. K. (1971). Frozen thin sections of fresh tissue for electron microscopy, with a description of pancreas and liver. *J. Cell Biol.* **51**, 772.

Civen, M., Ulrich, R., Trimmer, B. M., and Brown, C. B. (1967). Circadian rhythms of liver enzymes and their relationship to enzyme induction. *Science* **157**, 1563.

Clift, F. F., and Cook, R. P. (1932). A method of determination of biologically important aldehydes and ketones, with special reference to pyruvic acid and methylgloxal. *Biochem. J.* **26**, 1788.

Cope, G. H. (1968). Low-temperature embedding in water-miscible methacrylates after treatment with antifreezes. *J. Roy. Micros. Soc.* **88**, 235.

Cotran, R. S., and Karnovsky, M. J. (1968). Ultrastructural studies on the permeability of the mesothelium to horseradish peroxidase. *J. Cell Biol.* **37**, 123.

Dahl, H. A., and From, S. Hj. (1971). Some effects of polyvinyl alcohol and polyvinyl pyrrolidone on the activity of lactate dehydrogenase and its isoenzymes. *Histochemie* **25**, 182.

Dantzler, W. H. (1972). Effects of incubations in low potassium and sodium media on Na-K-ATPase activity in snake and chicken kidney slices. *Comp. Biochem. Physiol.* **41B**, 79.

Davis, J. M., and Himwich, W. A. (1971). The amino acid, norepinephrine, and serotonin content of rat brain fixed with glutaraldehyde. *Brain Res.* **33**, 568.

Deane, H. W. (1963). Nuclear localization of phosphatase activity: fact or artifact. *J. Histochem. Cytochem.* **11**, 443.

De Jong, D. W., Olson, A. C., and Jansen, E. F. (1967). Glutaraldehyde activation of nuclear phosphatase in cultured plant cells. *Science* **155**, 1672.

Desmet, V. J. (1962). The hazard of acid differentiation in Gomori's method for acid phosphatase. *Stain Technol.* **37**, 373.

Elford, B. C., and Walter, C. A. (1972). Effects of electrolyte composition and pH on the structure and function of smooth muscle cooled to $-79°C$ in unfrozen media. *Cryobiology* **9**, 82.

Ellar, D. J., Nunoz, E., and Salton, M. R. J. (1970). The effects of low concentrations of glutaraldehyde on *Microcossus lysodeikticus:* changes in the release of membrane-associated enzymes and membrane structure. *Biochim. Biophys. Acta* **225**, 140.

Epton, R., McLaren, J. V., and Thomas, T. H. (1971). Enzyme insolubilization with cross-linked polyacryloylaminoacetaldehyde dimethylacetal. *Proc. Biochem. Soc.* **123**, 21.

Ericsson, J. L. E. (1966). On the fine structural demonstration of glucose-6-phosphatase. *J. Histochem. Cytochem.* **14**, 301.

Ericsson, J. L. E., and Trump, B. F. (1965). Observations on the application to electron microscopy of the lead phosphate technique for the demonstration of acid phosphatase. *Histochemie* **4**, 470.

Ernst, S. A. (1972). Transport adenosine triphosphatase cytochemistry. II. Cytochemical localization of ouabain-sensitive, potassium-dependent phosphatase activity in the secretory epithelium of the avian salt gland. *J. Histochem. Cytochem.* **20**, 23.

Ernst, S. A., and Philpott, C. W. (1970). Preservation of Na-K-activated and Mg-activated adenosine triphosphatase activities of avian salt gland and teleost gill with formaldehyde as fixative. *J. Histochem. Cytochem.* **18**, 251.

Essner, E., Novikoff, A. B., and Masek, B. (1958). Adenosine triphosphatase and 5-nucleotidase activities in the plasma membrane of liver cells as revealed by electron microscopy. *J. Biophys. Biochem. Cytol.* **4**, 711.

Etherton, J. E., and Botham, C. M. (1970). Factors affecting lead capture methods for the fine localization of rat lung acid phosphatase. *Histochem. J.* **2**, 507.

Fahimi, H. D., and Drochmans, P. (1965). Essai de standardisation de la fixation au glutaraldéhyde. I. Purification et détermination de la concentration du glutaraldéhyde. *J. Microscopie* **4**, 725.

Farrant, J. (1970). Mechanisms of Injury and Protection in Living Cells and Tissues at Low Temperatures. In: *Current Trends in Cryobiology* (Smith, A. U., ed.). Plenum Press, New York and London.

Farrant, J. (1972). Human red cells under hypertonic conditions: a model system for investigating freezing damage. *Cryobiology* **9**, 131.

Fein, M. L., and Harris, E. H. (1962). *Quantitative Analytical Procedure for Determining Glutaraldehyde and Chrome in Tanning Solution.* U.S. Dept. Agri. Res. Serv. Pub. ARS-73-37.

Feuvray, D., and De Leiris, J. (1973). Effect of short dimethyl sulfoxide perfusions on ultrastructure of the isolated rat heart. *J. Mol. Cell. Card.* **5**, 63.

Forssmann, W. G., Siegrist, G., Orci, L., Girardier, L., Pictet, R., and Rouiller, C. (1967). Fixation par perfusion pour la microscopie électronique essai de généralisation. *J. Microscopie* **6**, 279.

Frederick, S. E., and Newcomb, E. H. (1969). Cytochemical localization of catalase in leaf microbodies (peroxisomes). *J. Cell Biol.* **43**, 343.

Frigerio, N. A., and Shaw, M. J. (1969). A simple method for determination of glutaraldehyde. *J. Histochem. Cytochem.* **17**, 176.

Gahan, P. B. (1965). Histochemical evidence for the presence of lysosome-like particles in the root meristem cells of *Vicia faba. J. Exp. Bot.* **16**, 350.

Gahan, P. B., Kalina, M., and Sharma, W. (1967). Freezing-sectioning of plant tissues: The technique and its use in plant histochemistry. *J. Exp. Bot.* **18**, 151.

Gander, E. S., and Moppert, J. M. (1969). Der Einfluss von Dimethylsulfoxide auf die Permeabilitat der Lysosomenmembrane bei quantitativer Darstellung der sauren Phosphatase. *Histochemie* **20**, 211.

Ghajar, B. M., and Harmon, S. A. (1968). The effect of dimethylsulfoxide (DMSO) on permeability of *Staphylococcus aureus. Biochem. Biophys. Res. Commun.* **32**, 940.

Ghosh, B. K. (1971). Grooves in the plasmalemma of *Saccharomyces cerevisiae* seen in glancing sections of double aldehyde-fixed cells. *J. Cell Biol.* **48**, 192.

Goldfischer, S., Essner, E., and Novikoff, A. B. (1963). The localization of phosphatase activities at the level of ultrastructure. *J. Histochem. Cytochem.* **12**, 72.

Goldfischer, S., Essner, E., Schiller, B. (1971). Nucleoside diphosphatase and thiamine pyrophosphatase activities in the endoplasmic reticulum and Golgi apparatus. *J. Histochem. Cytochem.* **19**, 349.

Göthlin, G., and Ericsson, J. L. E. (1971). Fine structural localization of acid phosphomonoesterase in the brush border region of osteoclats. *Histochemie* **28**, 337.

Hajós, F., and Kerpel-Fronius, S. (1970). The incubation of unfixed tissues for electron microscopic histochemistry. *Histochemie* **23**, 120.

Halperin, W. (1969). Ultrastructural localization of acid phosphatase in cultured cells of *Daucus carota. Planta* **88**, 91.

Hanker, J. S., Kusyk, C. J., Clapp, D. H., and Yates, P. E. (1970). Effect of dimethylsulfoxide (DMSO) on the histochemical demonstration of dehydrogenase. *J. Histochem. Cytochem.* **18**, 673.

Hayashi, S., Aramaki, Y., and Moguchi, T. (1972). Diurnal changes in ornithine decarboxylase activity of rat liver. *Biochem. Biophys. Res. Commun.* **46**, 795.

Hayat, M. A. (1970). *Principles and Techniques of Electron Microscopy: Biological Applications,* Vol. 1. Van Nostrand Reinhold Company, New York.

Hayat, M. A. (1972). *Basic Electron Microscopy Techniques.* Van Nostrand Reinhold Company, New York.

Heckman, C. A., and Barrnett, R. J. (1973). GACH: A water-miscible, lipid-retaining embedding polymer for electron microscopy. *J. Ultrastruct. Res.* **42**, 156.

Hilgertová, J., Straková, M., Vrbová, H., Gregorová, I., Šonka, J., and Josífko, M. (1972). Seasonal variations of human erythrocyte glucose-6-phosphate dehydrogenase activity in relation to temperature and dehydroepiandrosterone excretion. *Clin. Chim. Acta* **36**, 511.

Hinton, R. H., Burge, M. L. E., and Hartman, G. C. (1969). Sucrose interference in the assay of enzymes and proteins. *Analyt. Biochem.* **29**, 248.

Holt, S. J., and Withers, R. F. J. (1958). Studies in enzyme cutochemistry. V. An appraisal of indigogenic reactions for esterase localization. *Proc. Roy. Soc. B.* **148**, 520.

Hopsu-Havu, V. K., Arstila, V. K., Helminen, H. J., Kalimo, H. O., and Glenner, G. G. (1967). Improvements in the method for the electron microscopic localization of aryl sulfatase activity. *Histochemie* **8**, 54.

Hopwood, D. (1967a). Some aspects of fixation with glutaraldehyde: a biochemical and histochemical comparison of the effects of formaldehyde and glutaraldehyde fixation on various enzymes and glycogen, with a note on penetration of glutaraldehyde into liver. *J. Anat.* **101**, 83.

Hopwood, D. (1967b). The behavior of various glutaraldehydes on Sephadex G-10 and some implications for fixation. *Histochemie* **11**, 289.

Hopwood, D. (1972). Theoretical and practical aspects of glutaraldehyde fixation. *Histochem. J.* **4**, 267.

Hubbard, J. S., and Miller, A. B. (1969). Purification and reversible inactivation of the isocitrate dehydrogenase from an obligate halophile. *J. Bact.* **99**, 161.

Hündgen, M., Schäfer, D., and Weissenfels, N. (1971). Der Fixierungseinflus acht verschiedener Aldehyde auf die Ultrastrukur kultivierter Zellen. II. Der Strukturzustand des Cytoplasmas. *Cytobiologie* **3**, 202.

Janigan, D. T. (1964). The effects of aldehyde fixation on β-glucuronidase, β-galactosidase, N-acetyl-β-glucusaminidase, and β-glucosidase in tissue blocks. *Lab. Invest.* **13**, 1038.

Janigan, D. T. (1965). The effects of aldehyde fixation on acid phosphatase activity in tissue blocks. *J. Histochem. Cytochem.* **13**, 473.

Jansen, E. F., Tomimatsu, Y., and Olsen, A. C. (1971). Cross-linking of α-chymotrypsin and other proteins by reaction with glutaraldehyde. *Arch. Biochem. Biophys.* **144**, 394.

Kanamura, S. (1970). Difference in resistance to glutaraldehyde or formaldehyde fixation between mouse and rat hepatic glucose-6-phosphatase. *Acta Histochem. Cytochem.* **3**, 160.

Kanamura, S. (1971a). Demonstration of glucose-6-phosphatse activity in hepatocytes following transparenchymal perfusion fixation with glutaraldehyde. *J. Histochem. Cytochem.* **19**, 386.

Kanamura, S. (1971b). Ultrastructural localization of glucose-6-phosphatase activity in proximal convoluted tubule cells of rat kidney. *Histochemie* **28**, 288.

Karnovsky, M. J. (1965). A formaldehyde-glutaraldehyde fixative of high osmolarity for use in electron microscopy. *J. Cell Biol.* **27**, 137A.

Leduc, E. H., Bernhard, W., Holt, S. J., and Tranzer, J. P. (1967). Ultrathin frozen sections. *J. Cell Biol.* **34**, 773.

Lee, S. H., Dusek, J., and Rona, G. (1971). Electron microscopic cytochemical study of glutamic oxalacetic transaminase activity in ischemic myocardium. *J. Mol. Cell Cardiol.* **3**, 103.

Leskes, A., Siekevitz, P., and Palade, G. E. (1971). Differentiation of endoplasmic reticulum in hepatocytes. I. Glucose-6-phosphatase distribution *in situ. J. Cell Biol.* **49**, 264.

Lisý, V., Kovářů, H., and Lodin, Z. (1971). *In vitro* effects of polyvinylpyrrolidone and sucrose on the acetylcholinesterase, succinic dehydrogenase, and lactate dehydrogenase activity in the brain. *Histochemie* **26**, 205.

Löhr, G. W., and Walker, H. D. (1963). In: *Methods of Enzymatic Analyses* (Bermeyer, H. U., ed.). Academic Press, New York and London.

Lovelock, J. E., and Bishop, M. W. H. (1959). Prevention of freezing damage to living cells by dimethyl sulphoxide. *Nature* **183**, 1394.

McDowell, E. (1969). Observation on the ultrastructure and localization of acid phosphatase activity in the descending part of the proximal tubule of rat kidney. *J. Microscopie* **8**, 509.

McMillan, P. J., and Wittum, R. L. (1971). Lactic acid dehydrogenase isoenzymes of rat goleus muscle fibers as demonstrated by histochemical staining and electrophoresis. *J. Histochem. Cytochem.* **19**, 421.

Makita, T., and Sandborn, E. B. (1971). The effect of dimethyl sulfoxide (DMSO) in the incubation medium for the cytochemical localization of succinic dehydrogenase. *Histochemie* **26**, 305.

Malinin, T. (1972). Injury of human polymorphonuclear granulocytes frozen in the presence of cryoprotective agents. *Cryobiology* **9**, 123.

Manocha, S. L. (1970). Effect of glutaraldehyde fixation on the localization of various oxidative and hydrolytic enzymes in the brain of rhesus monkey, *Macaca mulatta. Histochem. J.* **2**, 249.

von, Matt, Ch. A., Fuenfschilling, H., Moppert, J. M., and Gander, E. S. (1971). Comparative determination of liver acid phosphatase activity in decapitated and perfused rats. *Histochemie* **25**, 72.

Meijer, A. E. F. H. (1972). Semipermeable membranes for improving the histochemical demonstration of enzyme activities in tissue sections. *Histochemie* **30**, 31.

Melnick, P. J. (1967). Histochemical enzyme procedures in viability assay of tissues and organs. *Cytobiology* **4**, 136.

Misch, D., and Misch, M. S. (1968). Lysosomes: histochemical demonstration of latency using dimethyl sulfoxide. *Third Int. Cong. Histochem. Cytochem.*, p. 179.

Morré, D. J., and Mollenhauer, H. H. (1969). Studies on the mechanisms of glutaraldehyde stabilization of cytomembranes. *Proc. Indiana Acad. Sci.* **78**, 167.

Moses, H. L., and Rosenthal, A. S. (1968). Pitfalls in the use of lead ion for histochemical localization of nucleoside phosphatases. *J. Histochem. Cytochem.* **16**, 530.

Nagata, T., and Murata, F. (1972). Supplemental studies on the method for electron microscopic demonstration of lipase in the pancreatic acinar cells of mice and rats. *Histochemie* **29**, 8.

Nir, I., and Seligman, A. M. (1971). Ultrastructural localization of oxidase activities in corn root tip cells with two new osmiophilic reagents compared to diaminobenzidine. *J. Histochem. Cytochem.* **19**, 611.

Novikoff, A. B. (1956). Preservation of fine structure of isolated liver cell particulates with polyvinylpyrrolidone-sucrose. *J. Biophys. Biochem. Cytol.* **2**, Suppl., 65.

Ogata, K., Ottesen, M., and Svendsen, I. (1968). Preparation of water-insoluble, enzymatically active derivatives of subtilisin type Novo by cross-linking with glutaraldehyde. *Biochim. Biophys. Acta* **159**, 403.

Ogawa, K., and Barrnett, R. J. (1965). Electron cytochemical studies of succinic dehydrogenase and dihydronicotinamide-adenine dinucleotide diaphorase activities. *J. Ultrastruct. Res.* **12**, 488.

Papadimitriou, J. M., and van Duijn, P. (1970). Effects of fixation and substrate protection on the isozymes of asparate aminotransferase studied in a quantitative cytochemical model system. *J. Cell Biol.* **47**, 71.

Pegg, D. E. (1970). Banking of cells, tissues, and organs at low temperatures. In: *Current Trends in Cryobiology* (Smith, A. U., ed.). Plenum Press, New York and London.

Pelletier, G., and Novikoff, A. B. (1972). Localization of phosphatase activities in the rat anterior pituitary gland. *J. Histochem. Cytochem.* **20**, 1.

Peracchia, C., Mittler, B. S., and Frenk, S. (1970). Improved fixation using hydroxyalkylperoxide. *J. Cell Biol.* **47**, 156a.

Peracchia, C., and Mittler, B. S. (1972a). Fixation by means of glutaraldehyde-hydrogen peroxide reaction product. *J. Cell Biol.* **53**, 234.

Peracchia, C., and Mittler, B. S. (1972b). New glutaraldehyde fixation procedures. *J. Ultrastruct. Res.* **39**, 57.

Peters, T., and Ashley, C. A. (1967). An artifact in autoradiography due to binding of free amino acids to tissues by fixatives. *J. Cell Biol.* **33**, 53.

Pugh, D. (1972). The fine localization of N-acetyl-β-glucosaminidase in rat tissues using an indoxyl substrate. *Ann. Histochim.* **17**, 55.

Quiocho, F. A., and Richards, F. M. (1966). The enzymatic behavior of carboxypeptidase-A in the solid state. *Biochemistry* **5**, 4062.

Reale, E., and Luciano, L. (1964). A probable source of errors in electron histochemistry. *J. Histochem. Cytochem.* **12**, 713.

Reiss, J. (1971). Dimethylsulfoxide as carrier in enzyme cytochemistry. *Histochemie* **26**, 93.

Riecken, E. O., Goebell, H., and Bode, Ch. (1969). Untersuchungen zum Einfluss von tiefen Temperaturen und Speicherdauer auf einige histochemisch nachweisbare Enzymaktivitaten in Leber, Niere und Jejunum der Ratte. *Histochemie* **20**, 225.

Sabatini, D. D., Bensch, K., and Barrnett, R. J. (1963). Cytochemistry and electron microscopy: The preservation of cellular ultrastructure and enzymatic activity by aldehyde fixation. *J. Cell Biol.* **17**, 19.

Sabatini, D. D., Miller, F., and Barrnett, R. J. (1964). Aldehyde fixation for morphological and enzyme histochemical studies with the electron microscope. *J. Histochem. Cytochem.* **12**, 57.

Sachs, D. H., and Winn, H. J. (1970). The use of glutaraldehyde as a coupling agent for ribonuclease and bovine serum albumin. *Immunochem.* **7**, 581.

Schäfer, D., and Hündgen, M. (1971). Der Einfluss acht verschiedener Aldehyde und des pH-Wertes auf die Dartstellbarkeit der Glukose-6-phosphatase in Kultivierten Zellen. *Histochemie* **26**, 362.

Schejter, A., and Bareli, A. (1970). Preparation and properties of cross-linked water-insoluble catalase. *Arch. Biochem. Biophys.* **136**, 325.

Seligman, A. M., Chauncey, H. H., and Nachlas, M. M. (1951). Effect of formalin on the activity of five enzymes of rat liver. *Stain Technol.* **26**, 19.

Seligman, A. M., Wasserkrug, H. L., and Plapinger, R. E. (1970). Comparison of the ultrastructural demonstration of cytochrome oxidase activity with three bis (phenylenediamines). *Histochemie* **23**, 63.

Sexton, R., Cronshaw, J., and Hall, J. L. (1971). A study of the biochemistry and cytochemical localization of β-glycerophosphatase activity in root tips of maize and pea. *Protoplasma* **73**, 417.

Sherman, J. K. (1972). Comparison of *in vitro* and *in situ* ultrastructural cryoinjury and cryoprotection of mitochondria. *Cryobiology* **9**, 112.

Shilkin, K. B., Papadimitriou, J. M., and Walters, M. N-I. (1971). The effect of dimethyl sulfoxide on hepatic cells of rats. *Aust. J. Exp. Biol. Med. Sci.* **44**, 581.

Shnitka, T. K., and Seligman, A. M. (1971). Ultrastructural localization of enzymes. *Ann. Rev. Biochem.* **40**, 375.

Smith, R. E., and Farquahar, M. G. (1966). Lysome function in the regulation of the secretory process in the cells of the anterior pituitary gland. *J. Cell Biol.* **31**, 319.

Smith, R. E., and Fishman, W. H. (1968). *p*-(Acetoxymercuric) aniline diazotate, a reagent for visualizing the Naphthol AS BI product of acid hydrolase action at the level of the light and electron microscope. *J. Histochem. Cytochem.* **17**, 1.

Sommer, J. R., and Hasselbach, W. (1967). The effect of glutaraldehyde and formaldehyde on the calcium pump of the sarcoplasmic reticulum. *J. Cell Biol.* **34**, 902.

Soriano, R. Z., and Love, R. (1971). Electron microscopic demonstration of acid phosphatase in nucleoli and nucleoplasm. *Exp. Cell Res.* **65**, 467.

Stuart, J., Bitensky, L., and Chayen, J. (1969). Quantitative enzyme cytochemistry of leukaemic cells. *J. Clin. Path.* **22**, 563.

Tennyson, V. M., Brzin, M., and Slotwiner, P. (1971). The appearance of acetylcholin-esterase in the myotome of the embryonic rabbit: an electron microscope cytochemical and biochemical study. *J. Cell Biol.* **51**, 703.

Torack, R. M. (1965). ATPase activity in rat brain following differential fixation with formaldehyde, glutaraldehyde, and hydroxyadipaldehyde. *J. Histochem. Cytochem.* **13**, 191.

Trump, B. F., and Bulger, R. E. (1966). New ultrastructural characteristics of cells fixed in a glutaraldehyde-osmium tetroxide mixture. *Lab. Invest.* **15**, 368.

Trump, B. F., and Ericsson, J. L. E. (1965). The effect of the fixative solution on the ultrastructure of cells and tissues: a comparative analysis with particular attention to the proximal convoluted tubule of the rat kidney. *Lab. Invest.* **14**, 1245.

Tyrer, J. H., Eadie, M. J., and Kukums, J. R. (1971). The post-mortem stability of certain oxidative enzymes in brain and spinal cord. *Histochemie* **27**, 21.

Van Harreveld, A., and Fifkova, E. (1972). Release of glutamate from the retina during glutaraldehyde fixation. *J. Neurochem.* **19**, 237.

Van Harreveld, A., and Khattab, F. I. (1968). Perfusion fixation with glutaraldehyde and postfixation with osmium tetroxide for electron microscopy. *J. Cell Biol.* **3**, 579.

Wang, J. H. C., and Tu, J. I. (1969). Modification of glycogen phosphorylase by glutaraldehyde. *Biochemistry* **8**, 4403.

Ward, P., and Smith, A. U. (1971). Preliminary studies of the effects of DMSO on histochemical reactions. *J. Microscopy* **94**, 139.

Weissenfels, N., Schäfer, D., and Hündgen, M. (1971). Der Fixierungseinflus acht verschiedener Aldehyde auf die Ultrastruktur kultivierter Zellen. I. Der Strukturzustand des Zellkerns. *Cytobiologie* **3**, 188.

Weston, P. D., and Avrameas, S. (1971). Proteins coupled to polyacrylamide beads using glutaraldehyde. *Biochem. Biophys. Res. Commun.* **45**, 1574.

Widnell, C. C. (1972). Cytochemical localization of 5′-nucleotidase in subcellular fractions isolated from rat liver. I. The origin of 5′-nucleotidase activity in microsomes. *J. Cell Biol.* **52**, 542.

Wise, G. E., and Flickinger, Ch. J. (1971). Patterns of cytochemical staining in Golgi apparatus of amebae following enucleation. *Expt. Cell Res.* **67**, 323.

2

Phosphatases

EDWARD ESSNER

Division of Cytology,
Sloan-Kettering Institute for Cancer Research,
New York

INTRODUCTION

During the past decade, staining methods for the localization of phosphatase activities have been applied extensively at the level of ultrastructure. Although these methods are relatively simple and easy to apply, they are subject to various artifacts which can limit their resolution or complicate the interpretation of the results; in some instances, there is doubt regarding the validity of the method itself. In the first part of this chapter, some general aspects of the phosphatase procedures are discussed; the second half of the chapter is devoted to a survey of specific methods, including composition of incubation media, that appear to be most useful for the demonstration of phosphatase activities.

The ultrastructural localization of phosphatase activities is based largely on the application of histochemical procedures involving the precipitation of metal salts, which were originally developed for light microscopy by Gomori (1939) and by Takamatsu (1939). These methods are based upon the theory that phosphate ions, liberated by enzymatic hydrolysis of various organic phosphates serving as substrates, will be trapped in *statu nascendi* (at site of formation) by metal cations present in the medium, to form highly insoluble precipitates. The precipitates of lead and calcium which are used most frequently as "trapping" ions, are colorless, but can be transformed, by exposure to sulfides, into blackish, highly insoluble reaction products.

Clearly, the degree of precision with which the end products of phosphatase activities can be localized within cells and tissues rests ultimately on the limits of resolution imposed by the light microscope. The adaptation of light microscopic staining methods to electron microscopy is an attempt to overcome this limitation and thereby establish the relationship between enzyme activity and cell ultrastructure. Although these comments refer specifically to phosphatases, they apply also to the localization of enzyme activities in general. To gain any advantage from the greater resolving power of the electron microscope, however, both tissue fine structure and enzymatic activity must be adequately preserved. Fine structure must resist the rigors imposed by incubation in chemically and osmotically active media, while enzyme activity must survive the inhibitory effects of fixation without diffusing from its sites of origin. The localization of an enzyme activity at the level of electron microscopy is always a compromise between these two sets of conflicting requirements. These problems have been treated in detail in several earlier reviews (Holt and Hicks, 1962; Pearse, 1963; Barrnett, 1964; Scarpelli and Kanczak, 1965) and in this volume and elsewhere by Hayat (1970).

In addition to the requirements discussed above, the end product (reaction product) of enzyme activity must also meet certain criteria: it should be highly insoluble and resistant to the solubilizing effects of dehydration and embedding media. In order to be detected in the electron microscope, the end product must have sufficiently high intrinsic electron scattering properties so that, when generated at active sites in adequate amounts, contrast will be enhanced over neighboring sites without obscuring cellular detail. According to Holt and Hicks (1962), an increase in mass of about 10% is sufficient to detect active sites regardless of the nature of the end product. The properties of metal-salt precipitates such as those generated in Gomori type media appear to satisfy some of these requirements in varying degrees. However, they have several disadvantages: the precipitates occur as discrete, fairly coarse particles which may obscure cell structures; they become continuous, only after relatively long incubations, as a result of the accumulation of excess product; and they tend to diffuse, particularly from sites of high activity, and to adsorb to other structures.

HISTORY

The first localizations of phosphatase activity at the electron microscope level were achieved by Sheldon and associates, who showed that the reaction products of acid and alkaline phosphatase were electron-opaque and could be detected at specific sites in the intestinal epithelial cells of the mouse (Sheldon *et al.*, 1955; Brandes *et al.*, 1956). In the case of acid phosphatase, the initial reaction product, lead phosphate, was converted into lead sulfide to enhance contrast. However, subsequent studies have shown that lead phosphate itself is sufficiently opaque to electrons and need not be converted to the corresponding sulfide. In these early studies, small blocks of tissue were fixed briefly in osmium tetroxide, then incubated in the phosphatase media, and processed in the usual manner for electron microscopy.

Although the above procedure was subsequently used to localize other phosphatase activities, it has been found unsuitable for general use because of the ease with which false localizations can occur. This is largely due to the relatively slow penetration of osmium tetroxide and to the fact that the tissue block is an impediment to the free diffusion of substrates and other reagents in the medium. Thus, when small blocks of tissue are exposed to osmium tetroxide for 1 to 2 hr, the outer few layers become well fixed, whereas the interior is relatively unfixed. However, enzyme activity in the well-fixed exterior may be rapidly inactivated; in the interior, unfixed enzyme may be free to diffuse to the periphery or into the medium, where it may hydrolyze the substrate and thus serve as a source of free phosphate.

The correct ratio of components in the Gomori medium is also critical, and variations in this ratio at different levels in the block can result in patterns of artifactual staining (Holt, 1959; Holt and Hicks, 1961b). Such deviations can occur as a consequence of unequal penetration of substrate and other reagents. The barrier to free diffusion appears to reside in the tissue block itself, rather than with the type of fixative, for even after formaldehyde (which penetrates more rapidly than glutaraldehyde) treatment, lead enters the block by unhindered diffusion, obeying Fick's law, while the substrate does not.

Several instances have been reported in which observations based on the use of tissue blocks could not be verified by light microscopy of formalin-fixed frozen sections in which diffusion artifact is minimized (Goldfischer *et al.*, 1964).

These considerations make it unlikely that the tissue block technique can be used for the general study of phosphatase localizations. However, this does not seem to apply to bone marrow or to pellets derived from blood or tissue culture cells, which are evidently more porous to the components of the incubation media.

In order to avoid diffusion artifacts in tissue blocks, Novikoff and colleagues suggested the use of 25 μ frozen sections of formol-calcium fixed tissue. Although this resulted in retention of sufficient enzyme activity, tissue fine structure was poorly preserved (Kaplan and Novikoff, 1959; Novikoff, 1959; Essner and Novikoff, 1961). In a detailed investigation of the conditions governing penetration of substrates and reagents into tissues, Holt and Hicks (1961b) showed that frozen sections of rat liver with a maximum thickness of 50 μ would allow rapid penetration of all components of the medium and result in uniform staining patterns. These authors also suggested the use of neutral 4%

formol-phosphate sucrose, previously mentioned by Wachtel *et al.* (1959) for improved preservation of fine structure (Holt and Hicks, 1961a).

Further significant improvements were obtained with the introduction of glutaraldehyde as the primary fixative (Sabatini *et al.*, 1963) and the substitution of more rigid epoxy resins such as Epon for methacrylate. Finally, the damage that occurs in preparing frozen sections, due to freezing and thawing the tissue, was considerably diminished with the introduction of instruments capable of chopping fixed tissue into slices as thin as 20 μ without freezing. At present, two such devices, employing different cutting principles, are commercially available (Smith, 1970).

The procedure in general use for demonstrating phosphatase activities for electron microscopy is indicated below. Light microscopy can serve as an important adjunct in assessing the electron microscope localizations. This can be done by incubating 10 μ formol-calcium fixed frozen sections along with those sections being prepared for electron microscopy. During the course of incubation, a quick mount of a frozen section is used to monitor the rapidity of the reaction; it also serves later as a guide to assessing the electron microscopic findings. Ordinarily, a light but detectable staining reaction observed in the frozen sections indicates that the incubation is of sufficient duration for electron microscopy. However, it is advisable to incubate the sections for electron microscopy for varying intervals of time, so that the most appropriate level of reaction product can be selected for study.

General Procedure for the Localization of Phosphatase Activities

1. Fix thin (1-2 mm) slices of tissue in buffered glutaraldehyde; rinse in buffer (several changes, 2 to 3 hr) or store in cold buffer overnight. Prepare 40 μ sections on a freezing microtone or cryostat. Collect sections in cold buffer or 5% sucrose solution.

Fix thin strips (1 × 5 mm) of tissue in buffered glutaraldehyde; rinse or store in buffer overnight as above. The strips can be conveniently prepared from thicker slices while still immersed in fixative. Chop strips into 40 μ sections with a Tissue Sectioner (Smith, 1970). Collect sections in cold buffer or 5% sucrose solution.

2. Incubate sections in phosphatase medium (containing 4 to 5% sucrose, if desired). Special care in preparing the phosphatase media may help eliminate non-specific precipitates. Precautions such as those taken by Hugon *et al.* (1970) in preparing glucose 6-phosphatase and thiamine pyrophosphatase media, which evidently result in cleaner preparations, appear applicable to other lead media as well. Szmigielski (1971) recommends the addition of dextran (M.W. 250,000) to prevent precipitation of lead. However, this procedure has not been tested extensively.

3. Rinse in buffer or sucrose solution. For the demonstration of acid phos-

phatase activity, some investigators employ a 1 min rinse in buffer containing sucrose and 4% formaldehyde (Miller and Palade, 1964).

4. If desired, visualize sections by immersion in dilute ammonium sulfide. Although not necessary for electron microscopy, this step is useful for identifying sites of phosphatase activity with light microscopy in semithin sections.

5. Fix sections in buffered (avoid phosphate buffers) 1% osmium tetroxide for 30 to 60 min. Prolonged osmication may dissolve lead phosphate or lead sulfide reaction products (Reale and Luciano, 1964).

6. If desired, treat sections with buffered uranyl acetate solution (Farquhar and Palade, 1965). Exposure should be limited to 1 hr (see discussion on diffusion).

7. Dehydrate and embed for electron microscopy. If chopped sections are used, it is advisable to embed the entire section and trim the block so as to include the free edges, where the reaction is more uniform and reproducible (Leskes *et al.,* 1971).

The above procedure can also be used for tissues such as bone marrow which can be fixed and incubated in suspension in the manner described by Bainton and Farquhar (1968). Tissue cultured cells growing in suspension (e.g., spinner cultures) can be handled in the same way, or the cells can first be pelleted, then fixed, cut into small fragments, and treated as indicated in the above procedure. Cells growing on cover slips should be fixed *in situ,* rinsed, and then incubated by floating the cover slips face down on the surface of the medium. After incubation, the cells are scraped off, collected by centrifugation, and prepared for electron microscopy. Other cover slips can be incubated and then mounted in glycerogel on a glass slide for light microscopic study.

SOURCES OF ARTIFACTS

Glutaraldehyde Fixation

As already mentioned, glutaraldehyde penetrates tissue more slowly than does formaldehyde. Thus, 40 μ frozen or chopped sections prepared from slices or even thin strips of glutaraldehyde-fixed tissue, may exhibit a well-fixed periphery and a poorly fixed inner core, from which soluble enzyme or enzyme reaction product may diffuse. According to Miller and Palade (1964), an unfixed core is not evident, at least in kidney, when 3 to 5 mm slices are fixed in 6.25% buffered glutaraldehyde for 2 to 3 hr.

Glutaraldehyde, even at relatively low concentrations, is a potent inhibitor of phosphatases. The degree of inhibition depends on the particular tissue and on the conditions of fixation; it may also be related to the purity of the glutaraldehyde (Anderson, 1967). Recent evidence suggests that the inactivation of acid phosphatase by glutaraldehyde is virtually complete within the first 5 min of fixation. Longer fixations may have a relatively minor inhibitory effect, but appear necessary in order to stabilize the tissue (Arborgh et al., 1971). Those phosphatase activities that persist to a degree sufficient for histochemical study include acid and alkaline phosphatase, glucose-6 phosphatase, and various nucleoside mono-, di-, and triphosphatases. It should be noted, however, that the persisting activity generally represents only a small fraction of the original activity; the localization of this "residual" activity is assumed to faithfully reflect the distribution of all enzyme present prior to fixation.

The degree of inhibition of phosphatase activity by glutaraldehyde may vary from one site to another in a particular tissue. This may be due to quantitative differences in the initial levels of activity, differences in the sensitivity to fixation which might reflect the differential inactivation of isozymes, or variations in the degree of fixation caused by unequal penetration of fixative. These observations again emphasize the necessity for assessing such localizations by light microscopy of formaldehyde-fixed tissues (Goldfischer et al., 1964).

At present, systematic studies on the effects of glutaraldehyde on phosphatase activities are available for only a few enzymes in selected tissues (Janigen, 1965; Anderson, 1967; Ericsson and Biberfeld, 1967; Hopwood, 1967; 1969).

Effects of Other Fixatives

Glutaraldehyde fixation has been employed most frequently for the electron microscopic localization of phosphatase activities; other fixatives, several in combination with glutaraldehyde, have been used less extensively. It should be noted that the degree of inhibition of phosphatase activities in a particular tissue may vary widely depending on the fixative used. This has been clearly shown in the case of brain tissue (Torack, 1965).

The fixative suggested by Karnovsky (1965), composed of glutaraldehyde and paraformaldehyde, penetrates tissues more rapidly than does glutaraldehyde alone, but is a stronger inhibitor of phosphatase activities. Therefore preliminary trials with dilutions of the original concentration are usually required to obtain satisfactory results. Hydroxyadipaldehyde is a weak fixative, and is not recommended for phosphatase localizations. Tissues treated with this fixative require inordinately long fixation to obtain satisfactory results. This is also borne out by the observations of Janigen (1965), who reported that a large proportion of the acid phosphatase activity in rat liver remains soluble following fixation with this agent.

Although not yet used extensively, paraformaldehyde alone apparently preserves labile enzymes such as mitochondrial ATPase activity (Ahrens and Weissenfels, 1969). Its potentialities as a fixative for phosphatases need to be explored more fully.

It is evident from this discussion that no ideal fixative or combination of fixatives is available for the localization of phosphatase activities. Glutaraldehyde remains the fixative of choice, provided the limitations discussed above are recognized and precautions are taken to minimize diffusion artifacts. Fixative combinations that enhance the penetration of glutaraldehyde into tissues deserve further study. Enzyme inhibition may be significantly reduced with such fixatives, since they seem to be effective even when relatively low concentrations of glutaraldehyde are employed.

Diffusion and Adsorption Artifacts

A disturbing feature of the Gomori lead salt method is the fact that localizations of enzyme reaction product at specific sites are invariably accompanied by scattered deposits elsewhere in the section. The possible sources of this kind of precipitate have already been discussed.

It is not yet clear, however, whether these deposits consist of lead, lead phosphate, or both. Gomori (1952) considered the deposits to be lead, and introduced the so-called acid rinse to remove them from frozen sections. There is little doubt that lead binding also occurs at the ultrastructural level. Recent studies have shown that the lead present in incubation media that lack substrate or contain an appropriate enzyme inhibitor can adsorb preferentially to organelles such as endoplasmic reticulum and mitochondria in a fashion that simulates enzyme localizations (Behnke, 1966; Vethamany and Lazarus, 1967; Podolsky, 1968; Wilson, 1969). For example, Gillis and Page (1967) were able to demonstrate lead binding to striated muscle fibers by first exposing them to an ATPase medium containing lead without substrate and then immersing the fibers in solutions of inorganic phosphate. Although lead per se was not observed in the muscle fibers, subsequent exposure to phosphate resulted in the deposition of lead phosphate in patterns that mimicked the enzyme localizations obtained in the complete medium; however, see Wills (1967).

An interesting method, which may prove useful in distinguishing heavy metal deposits due to adsorption from those generated by enzyme activity, is the use of color electron micrographs. According to Liptrap (1968), deposits of enzyme reaction product (lead phosphate) can be differentiated from deposits of lead citrate, lead hydroxide, or uranyl acetate by virtue of their differing densities, which are revealed by converting their black and white images into readily distinguishable colors.

According to Goldfischer *et al.* (1964), the random precipitate commonly observed after incubation in phosphatase media is not lead but lead phosphate, and is probably derived from the hydrolysis of substrate by soluble (unfixed) enzyme or by diffusion of lead phosphate from active sites. As these authors indicate, the acid rinse removes lead phosphate as well as lead, but does not distinguish between lead phosphate due to enzyme activity and that which is adsorbed secondarily to other sites as a result of diffusion. Unless the rinse is used under strictly controlled conditions, it may remove all the reaction product from sites having initially small accumulations. Caution is also advised in the use of buffered uranyl acetate solutions, which are commonly employed to enhance the contrast of membranes (Farquhar and Palade, 1966); this treatment also acts as an acid rinse, and can remove reaction product from enzyme sites (Seeman and Palade, 1967).

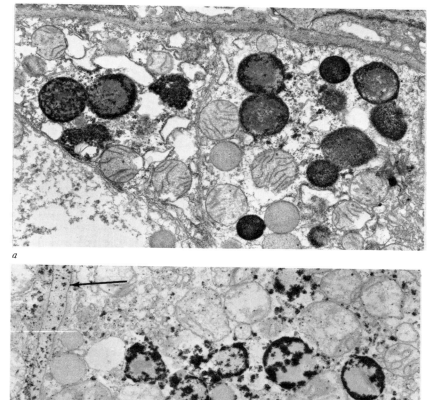

Fig. 2–1. Rat kidney incubated for acid phosphatase activity in Novikoff's CMP medium, pH 5.2. *a*: reaction product is localized to lysosomes. The limiting membranes have heavier accumulations, but there is relatively little diffusion. *b*: an area from the same thin section showing greater diffusion of reaction product around lysosomes and on nearby basement membrane (arrow). *a*, × 8,000; *b*, × 9,000.

Artifactually adsorbed precipitates, whether due to lead or to lead phosphate, are relatively sparse, and can often be distinguished from the heavier deposits of lead phosphate that accumulate at specific sites as a result of enzymatic activity. However, this does not necessarily apply to the diffusion that emanates from sites showing heavy accumulations of reaction product. In such instances, neighboring regions may also show heavy artifactual deposits of lead phosphate (Fig. 2–1b). Useful criteria for distinguishing diffusion artifact of this kind have been discussed by Novikoff et al. (1966).

LOCALIZATION IN ULTRATHIN SECTIONS

As already discussed, phosphatase activities are generally localized in tissues by incubating fixed frozen or nonfrozen sections in the appropriate medium; the sections are then fixed in osmium tetroxide, dehydrated, and embedded for electron microscopy. The disadvantages of this procedure, particularly as related to problems of diffusion, have been considered. An alternative approach is to embed tissue in a water-soluble medium and then to incubate thin sections of this material in phosphatase media. Since the penetration of reagents into thin sections is much more rapid and uniform, this technique has the advantage of greatly reducing diffusion artifacts. In an early study, alkaline phosphatase reaction product was localized in this manner by incubating thin sections of formalin fixed tissue that had been embedded in a water-soluble polyepoxide (Byczkowska-Smyk and Bernhard, 1960). This phosphatase has also been demonstrated in ultrathin sections prepared from frozen, dried intestine.

More recently, techniques for the routine preparation of ultrathin frozen sections have been developed, and this has led to renewed interest in their use for cytochemistry. Leduc et al. (1967) recently localized the reaction products of several phosphatase activities in ultrathin frozen sections of glutaraldehyde-fixed, gelatin-embedded kidney. In addition, a nucleoside phosphatase activity which, according to Vorbrodt and Bernhard (1968), is demonstrable in the nuclei has been localized with this technique. The methodology involved in preparing ultrathin frozen sections for routine use has been further improved (Bernhard and Viron, 1971; Iglesias et al., 1971), and several instruments designed for this purpose are now commercially available. It should be noted, however, that, in general, preservation of fine structure with these procedures is not as satisfactory as in tissues prepared in the conventional manner.

AZO DYE PROCEDURES

The difficulties inherent in the metal-salt precipitation methods have led some investigators to develop alternative methods that would be applicable to electron microscopy. Azo dye staining methods, although potentially very versatile, are limited by the propensity of the organic dye to diffuse into the embedding medium and by the relatively low contrast of the final pigment product. However, the opacity of the final product can be increased in several ways. For example, it has been shown that a heavy metal, firmly bound into a dye which serves as coupling agent, will become incorporated into the final dye product, thereby rendering it sufficiently opaque. This approach is preferable to the alternative of introducing metal ions into the preformed dye product by chelation, a process which usually results in a product soluble in both organic solvents and embedding media. Certain phosphatase activities have been demonstrated with electron microscopy by using heavy metal coupling agents as described above; two such procedures are described below in the section dealing with acid phosphatase.

Another approach to obtain increased contrast of the final dye product is based on the fact that certain dyes possess sulfur-containing groups capable of reacting with osmium tetroxide to form very opaque products of osmium black (Hanker et al., 1964; 1966). This method has been used for the localization of alkaline phosphatase activity. However, judging from the illustrations, the deposits of final product are relatively large and sparsely distributed, making precise localization difficult. It is of interest that certain azoindoxyl dyes of the type described by Holt and Hicks (1966) will also react with osmium tetroxide to form electron opaque complexes. These may prove useful in developing additional dye methods for the fine structural localization of phosphatase activities.

The various phosphatase media indicated below, under the appropriate headings, are those currently in use by most investigators. Incubations carried out with these media should be accompanied by controls in which substrate has been omitted. Inhibitors have not been included in this discussion, since their effects differ widely, depending upon the enzyme activity being demonstrated, the tissue, type of fixation, and other factors. However, most of the useful inhibitors can be found in the standard books on histochemistry or in biochemical reference works.

ACID PHOSPHATASE

The acid phosphatases (phosphomonoesterases II) consist of a group of nonspecific enzymes that hydrolyze a variety of organic esters with the liberation of phosphate ions. The pH optimum varies from $\sim$ 4.5 to 6.0. Incubations are usually carried out at pH 5.0 to 5.2.

Following the studies by Sheldon et al. (1955), the Gomori medium, which contains β-glycerophosphate as substrate, was used to localize acid phosphatase reaction product in many different types of cells. The first localizations of this enzyme in lysosomes were obtained by incubating formalin-fixed frozen sections (Essner and Novikoff, 1961; Holt and Hicks, 1961b) or small blocks of liver (Daems, 1962). Following these early observations, acid phosphatase activity was demonstrated in various types of lysosomes that are characteristic of different cell types as well as in other structures, including the endoplasmic reticulum and Golgi apparatus. A discussion of these localizations is beyond the scope of this chapter. A detailed discussion may be found in reviews by Novikoff (1963; 1967a); the latter reference deals mainly with lysosomes of neurons.

In addition to the nonspecific acid phosphatase which hydrolyzes β-glycerophosphate, Novikoff (1963) has identified a lysosomal "acid nucleotidase" (5′-nucleotidase) that hydrolyzes cytidine 5′ monophosphate (CMP) and other nucleoside monophosphates and is stimulated by Mn^{++} ions. It is not yet clear whether this activity is due to a different acid phosphatase or to another, more specific phosphatase. This activity should not be confused with the cobalt-stimulated nucleoside monophosphatase ($Co^{++}CMPase$) that has been demonstrated in plasma membranes (Novikoff, 1964). A 5′-nucleotidase activity hydrolyzing adenosine 5′-monophosphate at acid pH is also present in azurophile granules of leukocytes (Bainton and Farquhar, 1968).

In practice, CMP is often substituted for β-glycerophosphate in the Gomori medium. Since it is generally hydrolyzed more rapidly, incubation times can be shortened, thus reducing nonspecific precipitate. It is assumed that the localizations obtained with the two substrates are similar. However, in certain instances, differences in staining patterns have been noted, and it is therefore advisable to compare localizations obtained with the two substrates with light microscopy before CMP is used in lieu of glycerophosphate for electron microscopy.

As already mentioned, the absence of phosphatase activity at a particular site may be due to the inhibitory effects of glutaraldehyde. Another possibility, noted in the case of lysosomal acid phosphatase, is related to certain characteristics of the lead method. Instances have been recorded in both normal and experimentally altered tissue, in which acid phosphatase activity is not demonstrable in lysosomes with the lead medium, but is readily visualized with azo dye procedures such as that of Barka and Anderson (1962). The reasons for this discrepancy have been discussed by several authors (Rosenbaum and Rolon, 1962; Goldfischer et al., 1964; Seeman and Palade, 1967; Bainton and Farquhar, 1968).

A related problem is whether all the acid phosphatase activity in a given tissue can be revealed with the lead medium. According to some authors, acid phosphatase activity can be demonstrated in ribosomes and in endoplasmic reticulum, but only with the azo dye method, since the enzyme at these sites can hydrolyze naphthol derivatives but not glycerophosphate. However, the electron micrographs illustrating some of these findings are presented at relatively low magnifications, making the details of these localizations difficult to interpret (Livingston et al., 1969, Fig. 3).

When the reaction product of acid phosphatase activity is localized in lysosomes, the heaviest accumulations are usually found at the limiting membrane, and lighter deposits in the interior of the organelles (Fig. 2-1). This suggests that the reaction product originates at the membrane. However, since the product may have diffused from the interior, it is not possible to determine unequivocally the precise site of origin.

A modification of the Gomori medium designed to reduce diffusion artifact was introduced by Barka and Anderson (1962) (see below). This medium contains a lower concentration of lead together with maleate buffer, which is considered to stablize the lead ions and maintain them in solution. Although this modification is apparently preferred by many investigators, there is no clear evidence that it has any advantages over the original medium. In this author's laboratory, random precipitate appears to occur to approximately the same degree with both procedures.

A new method for the demonstration of acid phosphatase activity, recently reported in preliminary form by Fitzsimons et al. (1970), depends not on lead but on the formation of a complex of gold and adenosine monophosphate. Enzymatic hydrolysis at acid pH liberates the phosphate group while precipitating the alcohol-gold moiety, which is electron-opaque and is apparently deposited at sites of enzyme activity.

As already discussed, the azo dye methods have inherent difficulties that limit their usefulness for the localization of phosphatase activities at the electron microscope level. However, there are two methods which appear to be potentially useful; in both, the contrast of the final reaction product is increased by incorporating a heavy metal into the coupling agent, as discussed previously.

In the first method (Tice and Barrnett, 1965), the medium contains a naphthol phosphate as the substrate and lead phthalocyanin (tri-4-diazo phthalocyanin) as the capture reagent. When tested on kidney, the precipitated dye product was confined to lysosomes without evidence of diffusion to other structures. The opacity of the final product, although lower than that of lead phosphate, is apparently sufficient to be detected over the native density of the lysosomes. As previously noted, Livingston et al. (1969) suggest an improved procedure which they claim can be used also to localize acid phosphatase activities in sites other than lysosomes.

In the other dye method (Smith and Fishman, 1969), naphthol AS-BI phosphoric acid is used as substrate, and the coupling agent is diazotized acetoxymercuric aniline. Following incubation in this medium, the initially low contrast of the final dye product is enhanced by reacting it with thiocarbohydrazide, which forms an electron-opaque mercury-sulfur linkage (Hanker et al., 1966). However, as the authors indicate, the method is limited, since some of the increased opacity observed in the lysosomes is apparently not due to the dye product.

The original Gomori lead medium and the two modifications discussed above are given below. The dye methods are not included here since they contain rather lengthy procedures for preparing the coupling agents. The original publications may be consulted for the details of these methods. The components of all the lead media shown below should be added in the order indicated; add lead solutions dropwise with continuous stirring. Filter the final medium before use.

Gomori's lead medium (Gomori, 1952)

Na-β-glycerophosphate (0.01 M)	30 mg
0.05 M acetate buffer, pH 5.0	11 ml
12% lead nitrate (final concentration 3.3 mM)	0.1 ml

Preincubate medium for 1 hr at 37°C; overnight incubation at 37°C has also been recommended (Ericsson and Trump, 1965).

Barka and Anderson's medium (Barka and Anderson, 1962)

Na-β-glycerophosphate (adjust to pH 5.0 with 1N HCl)	12.5 mg
Distilled water	1.0 ml
0.2 M tris-maleate buffer, pH 5.0	2.0 ml
0.2% lead nitrate (final concentration 2.4 mM)	2.0 ml

Novikoff's CMP medium (Novikoff, 1963)

Cytidine 5' monophosphate (CMP)	25.0 mg
Distilled water	12.0 ml
0.05 M acetate buffer, pH 5.0	10.0 ml
1% lead nitrate (final concentration 3.6 mM)	3.0 ml

ALKALINE PHOSPHATASE

As in the case of the acid phosphatases, the alkaline phosphatases (phosphomonoesterases I) are nonspecific enzymes that hydrolyze virtually all monoesters of phosphoric acid with the liberation of orthophosphate. These enzymes are active on the alkaline side of neutrality, especially at pH 9.0 and above.

The procedures for the localization of alkaline phosphatase reaction product are generally based on either the direct or the indirect (two-step) precipitation of a heavy metal such as calcium, cobalt, silver, lead, or cadmium. In the original method, devised for light microscopy, calcium is used at alkaline pH to trap the phosphate liberated from the substrate by enzymatic hydrolysis (Gomori, 1952). The resulting primary reaction product, calcium phosphate, is converted in a second step into cobalt phosphate and finally into cobalt sulfide. Calcium phosphate has only moderate opacity to electrons, but it can be increased, as shown by Brandes *et al.* (1956), by converting the precipitate into a silver salt.

The same result is achieved by converting calcium phosphate into a lead salt (Molnar, 1952; de Thé, 1965). This can be accomplished simply by treating the incubated sections with 2% lead nitrate for a few minutes (Wetzel *et al.,* 1967). In addition to the increased opacity, one advantage of the lead salts as final products is that, unlike salts of cobalt, they are relatively resistant to postfixation in osmium tetroxide (Reale and Luciano, 1964).

Although the conversion of calcium phosphate into salts of heavier metals appears to be a simple and adequate expedient for obtaining increased opacity, the additional step theoretically increases the risk of diffusion. For this reason, some early investigators have preferred to rely on the moderate opacity of calcium phosphate rather than risk diffusion from a subsequent conversion (Reale, 1962; Chase, 1963). However, it should be noted that in some of these studies false localizations apparently occurred as a result of the use of tissue blocks (Goldfischer et al., 1964).

The problem of low opacity of the primary reaction product as well as diffusion from secondary conversions can be avoided by using heavy metals in direct, rather than indirect, reactions. For example, in the method of Mölbert et al. (1960a and b) and von Deimling (1964), lead nitrate was used as the capture reagent in a direct reaction. The chelating agent, potassium sodium tartrate, was added to stabilize lead ions, and the reaction was carried out at a relatively low pH. However, this method has been criticized by several authors on the grounds that at a lower pH enzymes other than alkaline phosphatase may be active (Hugon and Borgers, 1966; Mizutani, 1966). In addition, tissue blocks in which diffusion may have occurred were used in these studies.

However, it is of interest that when frozen sections, rather than blocks, are incubated in the Mölbert medium, the results are comparable to those obtained with the Gomori medium (Goldfischer et al., 1964). According to Mizutani (1966), the amount of tartrate in the Mölbert medium is excessive; at lower concentrations, he was able to demonstrate alkaline phosphatase activity with this medium at an alkaline pH. A method based on the use of cadmium has also been described. However, it appears to be useful in kidney and endothelium of certain tissues but not in intestine (Mizutani and Barrnett, 1965).

Additional direct lead methods that appear to be useful for the localization of alkaline phosphatase activity are those of Hugon and Borgers (1966) and of Mayahara et al. (1967; see below). In the former, lead nitrate is used without a chelating agent by taking advantage of the fact that the solubility of lead ion is enhanced in maleate buffer (Barka and Anderson, 1962). Nevertheless, a precipitate, probably lead hydroxide, rapidly forms in the medium; the procedure is thus limited to sites of high activity that can be demonstrated with very brief incubations (Hugon, 1970). The latter medium is based on the earlier work of Tranzer (1965), who substituted lead citrate for lead nitrate. However, the concentration of lead (2 mM) in this case is considerably lower than in Tranzer's medium (12 mM), and undoubtedly aids in reducing enzyme inhibition due to lead as well as the adsorption of lead to the section. It should be noted that the substrate concentration in this medium appears to be inordinately high. The reader is referred to the original reference for details of this method.

All the media discussed above contain β-glycerophosphate as substrate. According to Oledzka-Slolwinska et al. (1967), CMP can also be used to demonstrate "alkaline phosphatase" activity. However, the authors admit that at least part of the activity observed in liver and kidney may have been due to a specific nucleoside monophosphatase.

Gomori's medium (Gomori, 1952)

β-Glycerophosphate	100 mg
Distilled water	8.5 ml
0.2 M 2-amino-2-methyl-1,3,propandiol buffer (pH 9.4)	10 ml
0.1 M Calcium chloride	1 ml
2% magnesium chloride	0.5 ml

Filter before use.
After incubation, treat sections briefly with 2% lead nitrate (Wetzel *et al.*, 1967).

Mizutani's medium (Mizutani, 1966)

Na-β-glycerophosphate	25 mg
Distilled water or 0.1 M $MgSO_4$ (1 ml) and water (1.7 ml)	2.7 ml
0.3 M 2-amino-2-methyl,3,propandiol buffer (pH 9.1)	3 ml
2% K-Na-tartrate	2.5 ml
1% lead nitrate (final concentration 5.1 mM)	1.3 ml

Adjust final medium to pH 9.2 with NaOH.

Hugon and Borgers' medium (Hugon and Borgers, 1966)

	pH 9.0	*pH 8.2*
Na-β-glycerophosphate	25 mg	25 mg
Distilled water	6.7 ml	7.7 ml
0.2 M Tris-maleate buffer (pH 9.0 or 8.2)	2.0 ml	1.0 ml
1% lead nitrate	1.3 ml	1.3 ml
Final lead concentrations	3.9 mM	3.6 mM

Adjust the pH with NaOH. Warm the medium at 37°C for 15 min, and keep at room temperature for 1 hr. Filter and use immediately.

GLUCOSE-6-PHOSPHATASE

Glucose-6-phosphatase is a specific enzyme capable of dephosphorylating glucose-6-phosphate as well as several other substrates. This activity was first demonstrated with light microscopy in liver by Chiquoine (1953; 1955), using a Gomori type lead medium; the medium was subsequently modified by Wachstein and Meisel (1956), who substituted Tris-maleate buffer and the potassium salt of glucose-6-phosphate.

Glucose-6-phosphatase activity is sensitive to fixatives, particularly formaldehyde, which can cause rapid inactivation of activity (Allen, 1961). Although
initially reported as being inactivated by glutaraldehyde fixation, the enzyme has
since been demonstrated in several studies employing this dialdehyde (Goldfischer *et al.,* 1964). Ericsson (1966) compared the preservation of hepatic fine
structure and of glucose-6-phosphatase activity after perfusion with glutaraldehyde, paraformaldehyde, and hydroxyadipaldehyde, and concluded that brief
perfusion with glutaraldehyde (3%, 1 to 2 min) gave the most satisfactory
results. Similar findings have been reported by Leskes *et al.* (1971) for fetal and
newborn liver. These authors also added substrate to the fixative in an attempt
to preserve enzyme activity. Kanamura (1971) has also reported good results
with transparenchymal perfusion of the liver, a technique that appears suitable
for embryos and very small animals.

Generally, no staining is observed with light microscopy when glucose-6-
phosphate is replaced by glucose-1-phosphate, fructose-1, 6-diphosphate, fruc-
tose-6-phosphate, ribose-5-phosphate, or β-glycerophosphate. Therefore these
substrates are frequently used as controls (Tice and Barrnett, 1962). Glucose-6-
phosphatase activity is also inhibited by Zn, Cu, Mg, and Fe (Chiquoine, 1953 &
1955; Tice and Barrnett, 1962). Thus, these ions can also serve as controls when
added to the complete medium.

The electron-microscopic localization of glucose-6-phosphatase activity was
reported by Tice and Barrnett (1962), who demonstrated reaction product in
the endoplasmic reticulum and nuclear envelope of hepatocytes, using blocks
and frozen sections of unfixed and hydroxyadipaldehyde-fixed liver. A similar
activity, whose function is not clear, has been identified in the endoplasmic
reticulum and nuclear envelope of pancreatic cells (Lazarus and Barden, 1964b),
blood and bone marrow cells (Rosen, 1969), and mouse jejunal epithelium
(Hugon *et al.,* 1970). According to Leskes *et al.* (1971), the concentration of
lead usually employed (see below) can be doubled to insure trapping of phosphate without any increase in enzyme inhibition.

When the reaction product of glucose-6-phosphatase activity is localized in
the endoplasmic reticulum and nuclear envelope, the initial deposits appear
within the cisternae and on the limiting membranes. With longer incubations, the
product tends to accumulate within the cisternae, and may sometimes obscure
the membranes. These observations suggest that the product originates from the
membranes. However, as discussed in the case of acid phosphatase, the possibility also exists that the reaction product moves from the lumen of the
cisternae to the limiting membranes.

Another important enzyme in carbohydrate metabolism that has been localized with electron microscopy is fructose 1,6-diphosphatase (Saito and Ogawa,
1968). This enzyme, which survives glutaraldehyde fixation, has been demonstrated by using lead citrate as trapping agent at high pH. The reaction product
appears as a very fine precipitate in the cytoplasmic matrix of hepatocytes. The

reaction is negative in fasting animals. These observations are consistent with the biochemical evidence that fructose 1,6-diphosphatase is a soluble enzyme. The original publication may be consulted for details of the method.

Glucose-6-phosphatase medium (Wachstein and Meisel, 1956)

Glucose-6-phosphatase	25 mg
Distilled water	27 ml
0.3 M Tris-maleate buffer, pH 9.7	20 ml
2% lead nitrate (final concentration 4.9 mM)	3 ml

NUCLEOSIDE PHOSPHATASES

These enzymes hydrolyze nucleoside mono-, di-, or triphosphates with release of the terminal phosphate. The nucleosides consist of a purine or pyrimidine base together with ribose. As mentioned previously, nucleoside monophosphatase (CMPase) activity has been used as a marker for lysosomes. However, it is the nucleoside di- and triphosphatases that have been studied most extensively by cytochemists. A very brief survey of the localization of these enzyme activities is given below. The localization of nucleoside phosphatase activities in the cytomembranes of various cell types has been reviewed in detail (Novikoff *et al.,* 1962).

Plasma Membranes

The nucleoside triphosphatase activity of primary concern to cytochemists is adenosine triphosphatase (ATPase). The localization of this enzyme is generally carried out with the Wachstein-Meisel lead medium or the calcium medium described by Padykula and Herman (1955); both media contain ATP as the substrate and Mg^{++} as the activating ion. This activity has been demonstrated in the plasma membranes of a variety of cell types. However, in many tissues, the enzyme activity in the plasma membranes exhibits broader specificity, often hydrolyzing all nucleoside triphosphates and, in a few instances (e.g., in certain neurons), nucleoside diphosphates as well. In these circumstances, the term "nucleoside phosphatase" (NDPase) activity is frequently used to designate this wider specificity. Nucleoside phosphatase activity has also been localized in the sarcoplasmic reticulum of cardiac and striated muscle cells, where it appears to be involved in the "calcium pump" mechanism.

The exact relationship between the Mg-stimulated ATPase demonstrable cytochemically and the so-called membrane (transport) ATPase identified biochemically, which participates in the active transport of cations (Skou, 1962), is not yet clear. Unlike transport ATPase, the cytochemically demonstrable activity (or activities) is not stimulated by cations or inhibited by the cardiac glycoside, ouabain; as already noted, it also possesses wider substrate specificity.

In addition, it has been argued that the transport ATPase cannot be demonstrated with the Wachstein-Meisel medium, since it is totally inhibited by lead at the concentrations normally employed in this medium. However, the possibility remains that the activity demonstrated by staining methods is related to a Mg-ATPase activity that does not function in active transport or in some other, as yet unknown, activity (Novikoff *et al.,* 1962; Novikoff, 1967a). This problem has been considered in detail by Tormey (1966), Farquhar and Palade (1966), and Marchesi and Palade (1967).

The first electron-microscopic localizations of ATPase activity were obtained by incubating osmium-fixed tissue blocks (Essner, *et al.,* 1958; Persijn *et al.,* 1961) or formol-calcium-fixed frozen sections (Kaplan and Novikoff, 1959) in the Wachstein-Meisel lead medium. This activity has since been demonstrated with electron microscopy in many tissues, including those that are specialized for active transport.

The precise localization of the ATPase reaction product in plasma membranes has not been conclusively established. In most instances, the deposits occur on the free surface of plasma membranes, and not on the inner or cytoplasmic side. In isolated cells, the product also accumulates at the free surface, and may diffuse from this site into the medium. In many tissues, accumulation of the reaction product occurs within the space between plasma membranes of adjacent cells (Fig. 2–2). At specialized surfaces, such as the bile canalicular membranes of hepatocytes, the reaction product is initially deposited on the free surface of the microvilli that project into the lumen; after longer incubation, it accumulates within the lumen.

However, it is interesting to note that when plasma membrane preparations (ghosts) derived from erythrocytes are incubated for ATPase activity, the reaction product is consistently found at the inner, rather than outer, face of the membrane (Marchesi and Palade, 1967). This problem might be approached more profitably using labeled antibody techniques, which are much less subject to diffusion.

Mitochondrial ATPase

Mitochondrial ATPase activity has been demonstrated with light microscopy using the Wachstein-Meisel (1957) and Padykula and Herman (1955) media. This activity is particularly sensitive to glutaraldehyde fixation, but can be demonstrated after formaldehyde or paraformaldehyde fixation. The enzyme activity demonstrated cytochemically is thought to correspond to the Mg^{++}-dependent, dinitrophenol-stimulated ATPase that has been studied extensively in mitochondria by biochemical methods (Pullman *et al.,* 1960). However, the exact relationship is uncertain, since the cytochemical activity is demonstrable with adenosine diphosphate and inosine diphosphate as well as with ATP (Rechardt and Kokko, 1967).

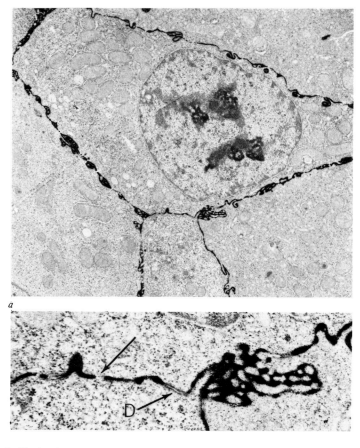

Fig. 2–2. Nucleoside phosphatase activity in a well-differentiated rat hepatoma; incubated in Wachstein-Meisel medium, pH 7.3 with ATP as substrate. *a*: reaction product has accumulated in space between opposing plasma membranes of tumor cells. *b*: higher magnification of an area from *a*, showing absence of product at points along membrane (arrow) and at desmosome-like structures (D). *a*, × 4,000; *b*, × 16,000.

Although the ATPase reaction product has been demonstrated in mitochondria with electron microscopy, the results have been contradictory (Ashworth *et al.*, 1963; Lazarus and Barden, 1964a; Essner *et al.*, 1965; Vethamany and Lazarus, 1967; Ogawa and Mayahara, 1969; Ahrens and Weissenfells, 1969). Some authors have found the product within the space between the membranes of individual cristae; other observations place the deposits within the mitochondrial matrix, sometimes in juxtaposition to the outer membrane of the cristae. This latter view, according to Grossman and Heitkamp (1968), is consistent with the biochemical evidence that ATPase is associated with the elementary particles coating the outer membrane of the cristae facing the matrix. However, it is uncertain whether the lead salt precipitates (well known for their propensity to diffuse) can distinguish active sites with this degree of precision.

The Wachstein-Meisel medium, indicated below, can also be used for the demonstration of certain other nucleoside phosphatase activities particularly mono- and triphosphatases by substituting the appropriate substrate for ATP. When demonstrating mitochondrial ATPase, some authors add 2,4 dinitrophenol (0.1 mM to 5.0 mM) to the complete medium.

Adenosine Triphosphatase medium (Wachstein and Meisel, 1957)

Adenosine triphosphate (Na)	25 mg
Distilled water	22 ml
0.2M Tris-maleate buffer, pH 7.2	20 ml
1.2% anhydrous magnesium sulfate	5 ml
2% lead nitrate (final concentration 3.6 mM)	3 ml

Lead-Catalyzed Hydrolysis of Nucleoside Phosphates

The conditions governing the formation and precipitation of lead phosphate in the Wachstein-Meisel medium and its subsequent partition between tissue and medium are complex and not completely understood (Berg, 1960). Lead inhibits the ATPase activity of unfixed tissue and, although the data are conflicting, it is apparently also capable of inactivating the activity of fixed tissue as well. Thus, the relatively small amount of residual ATPase activity that survives fixation is further reduced by the inhibitory effects of lead. The effective concentration of lead ions in the medium is also dependent on the other components present. For example, it has been shown that ATP can chelate lead ions, thereby decreasing the amount available either to precipitate the phosphate released enzymatically or to inhibit enzyme activity (Berg, 1964; Tice, 1969). The activating ions in the medium also apparently influence the concentration of lead. Obviously, any study involving the lead-ATPase medium must take into account the consequences of these various interactions.

In 1966, Rosenthal and co-workers initiated a series of studies which drew attention to another hitherto undescribed property of lead: its ability to catalyze a nonenzymatic, dephosphorylation of ATP and other nucleoside phosphates (Rosenthal *et al.*, 1966; Moses *et al.*, 1966). The extent of this hydrolysis depends on the relative concentrations of lead and ATP as well as on the presence of other divalent cations (Tice, 1969). According to Moses *et al.* (1966), the reaction represents a significant source of free phosphate. For example, in a reaction mixture (without tissue), containing 0.72 mM Na ATP, 0.36 mM lead nitrate (or greater), and Tris-maleate buffer, pH 7.2, the lead-induced hydrolysis of ATP results in the liberation of 0.3 μ moles phosphate/ml reaction mixture/90 min of incubation. This amount of phosphate is apparently sufficient to account for part or all of the reaction product deposited in plasma membranes, which is ordinarily attributed to enzymatic activity.

It has further been shown that the lead-induced hydrolysis of ATP occurs under the same conditions of incubation as those used to demonstrate the staining of plasma membranes. Indeed, if the concentration of lead is lowered to a point sufficient to suppress the lead-catalyzed reaction, the staining of plasma membranes is likewise diminished and nuclear and mitochondrial staining appear.

These and other observations have led to the conclusion that the plasma membrane localizations seen with the Wachstein-Meisel medium may not indicate sites of ATPase activity but, rather, selective affinity of phosphate or compounds of phosphate, presumably derived from the lead-catalyzed hydrolysis of ATP, for tissue reactive sites (Rosenthal et al., 1966; Moses et al., 1966; Moses and Rosenthal, 1967; 1968; Ganote et al., 1969; Rosenthal et al., 1969a and b; Rosenthal et al., 1970).

On the other hand, Novikoff (1967b; 1970) and others have marshaled considerable evidence supporting the validity of the ATPase procedure. Novikoff has stressed the fact that strong plasma membrane staining occurs at 4°C (5 to 10 min), at which point the degree of lead-catalyzed hydrolysis of ATP is insignificant (Rosenthal et al., 1966; Novikoff, 1970). Moreover, it has been shown that although lead catalyzes the hydrolysis of both ATP and ADP to the same extent, plasma membrane staining is frequently demonstrable with ATP, but not with ADP. It should also be noted that plasma membranes are reactive in the medium of Padykula and Herman (1955), which contains calcium, not lead.

Perhaps most compelling, however, is the evidence derived from observations of kidney. In this organ, the plasma membranes at the base of the tubule cells form deep infoldings in the cytoplasm and are often situated close to elements of the endoplasmic reticulum. With ATP as substrate, the reaction product is clearly localized to the invaginated plasma membranes, and is not found in the endoplasmic reticulum; conversely, with IDP as substrate, the product is present in the endoplasmic reticulum, and is totally absent from the plasma membranes (Novikoff, 1967b). This example serves to emphasize the high degree of substrate specificity exhibited by enzyme activities in different types of membranes, and thus strengthens the validity of the staining method.

Two additional points raised by Novikoff (1967b) in support of the validity of the method are: lead phosphate formed by slowly adding inorganic phosphate to the Wachstein-Meisel medium does not result in artifactual adsorption; plasma membrane staining is suppressed by p-chloromercuribenzoate (p-CMB), an inhibitor of ATPase (see also Wills, 1967). The inhibitor studies have been criticized by Moses and Rosenthal (1967) on the grounds that the effects of p-CMB on the nonenzymatic hydrolysis of ATP were not investigated. However, Mietkiewski et al. (1970) have recently shown that a number of inhibitors of ATPase, including p-CMB, have little or no effect on this reaction.

At present, there appears to be general agreement that although the lead-catalyzed hydrolysis of ATP does occur in the Wachstein-Meisel medium, this reaction alone cannot account for all the reaction product seen in cytochemical preparations. Indeed, the extent to which this reaction occurs at all in fixed tissue is still debatable. Thus, Jacobsen and Jorgensen (1969) have repeated the substantive experiments of Rosenthal and co-workers, and concluded that the contribution of lead-induced hydrolysis of ATP to the staining of fixed kidney tissue is negligible.

Apart from the question of the extent of lead-catalyzed ATP hydrolysis, there is the problem of the mechanisms involved in the generation of phosphate in this reaction. Moses and Rosenthal (1967) suggest that lead-induced ATP hydrolysis may involve transphosphorylation reactions of the type described by Lowenstein and co-workers (Lowenstein, 1958; Tetas and Lowenstein, 1963).

Obviously, simple nonspecific adsorption of lead phosphate formed in such reactions could not account for the marked specificities observed in different membranes. It has been shown, however, that the reaction product does not consist of lead phosphate alone, but also contains significant amounts of nucleotide; thus, the required specificity might reside in the differential affinity of tissue sites for the different lead phosphate-nucleotide complexes. At present, however, there is no substantive evidence that tissue sites, in particular plasma membranes or other types of membranes, can discriminate between nucleotide complexes of this kind with the necessary degree of selectivity.

NUCLEOSIDE DIPHOSPHATASE
AND THIAMINE PYROPHOSPHATASE

The localization of nucleoside diphosphatase (NDPase) activity in the endoplasmic reticulum and Golgi apparatus was first demonstrated by Novikoff and co-workers using a modification of the Wachstein-Meisel medium (Novikoff and Goldfischer, 1961; Novikoff et al., 1962).

In general, NDPase activity can be demonstrated by using as substrates the diphosphates of guanosine, uridine, or inosine, which in certain cell types are all rapidly hydrolyzed by endoplasmic reticulum and Golgi apparatus. Inosine diphosphate is usually employed in such studies and inosine diphosphatase (IDPase) activity is considered a useful "marker" for the endoplasmic reticulum.

The function of NDPase activity is not known, but it is of interest that the relatively few cell types having this activity also possess glucuronyl transferase, an enzyme involved in the biosynthesis of bilirubin glucuronide, suggesting that the two may be functionally related.

In addition to hydrolyzing the nucleoside diphosphates mentioned above, the Golgi apparatus of many cells will also split thiamine pyrophosphate (TTP); the resulting activity (TPPase), is often used as a marker for this organelle. TPP resembles the nucleoside diphosphates in having a pyrimidine base, but differs from them in having thiazole instead of ribose. Generally, cells that possess NDP-ase activity in the endoplasmic reticulum and nuclear envelope also have low levels of thiamine pyrophosphatase in these sites (Fig. 2–3a).

In liver, where they have been studied in detail, NDPase and TPPase activities in hepatocytes are similar but not identical when compared cytochemically; they differ not only in intracellular distribution but also in pH optima and in response to the addition of ATP (Goldfischer et al., 1971). It is not yet clear whether they represent one or several enzymes (Novikoff et al., 1962; Hündgen,

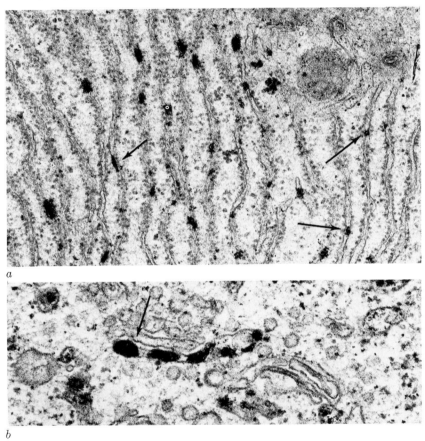

a

b

Fig. 2–3. a: Rat liver incubated in Novikoff and Goldfischer's TPP medium, pH 8.5. Enzyme reaction product is present within cisternae (long arrows) and on membranes (short arrow) of rough endoplasmic reticulum. Larger deposits that occlude the cisternae and mask the membranes are also evident (upper area of micrograph). b: Morris hepatoma, 5123 t.c., incubated in same medium with IDP as substrate, pH 7.0. Reaction product is localized within cavity of a Golgi saccule. No product is present, as a result of diffusion or activity, in narrow space between saccules or in the adjacent saccule despite close proximity of adjacent saccule membranes. a, × 48,000; b, × 56,000.

1970). Yamazaki and Hayaishi (1968), who recently studied a highly purified enzyme preparation from liver, have concluded that one enzyme with different pH optima is responsible for the observed activities. A nucleolar reaction with these substrates has also been described (Novikoff and Goldfischer, 1961).

When the IDPase reaction product is localized by electron microscopy in the endoplasmic reticulum (Fig. 2–3a) and nuclear envelope, the distribution of precipitate is virtually identical to that of glucose-6-phosphatase. In the Golgi apparatus, reaction products of NDPase and TPPase activities are found within the cavities of the saccules (Fig. 2–3b). Generally, little precipitate is found in the space between saccules, suggesting that significant diffusion from one saccule to another does not occur; possibly the limiting membrane of the saccules serves as an effective barrier to the movement of reaction product. The lead method, in this instance, can distinguish between reactive and nonreactive saccules, even when the two structures are in close apposition (Fig. 2–3b).

Nucleoside diphosphatase medium (Novikoff and Goldfischer, 1961)

Inosine diphosphate (Na) or thiamine pyrophosphate (Cl)	25 mg
Distilled water	7 ml
0.2 M Tris-maleate buffer, pH 7.2	10 ml
0.5% manganeses chloride	5 ml
1% lead nitrate (final concentration 3.6 mM)	3 ml

CYCLIC NUCLEOTIDE PHOSPHODIESTERASE

Cyclic adenosine 3′,5′ monophosphate (cyclic AMP) plays a fundamental role in mediating cellular responses to certain hormones and other agents (Sutherland and Rall, 1960). The levels of cyclic AMP in tissues are, in turn, regulated by the enzyme phosphodiesterase (cyclic 3′5′ nucleotide phosphodiesterase), which catalyzes the hydrolysis of cyclic AMP to adenosine 5′ monophosphate (5′ AMP) (Cheung, 1970). Although phosphodiesterase is not strictly a phosphatase, it is included in this discussion in view of its important relationship to cyclic AMP.

Shanta *et al.* (1966) devised a light microscopic method for the demonstration of phosphodiesterase activity based on a modification of a typical biochemical assay system. They reported the localization of activity in cryostat sections of unfixed, frozen brain, and other tissues. Their medium contains cyclic AMP as substrate, lead acetate, and snake venom as a source of 5′ nucleotidase. The hydrolysis of cyclic AMP results in the formation of 5′ AMP, which, in turn, is converted to adenosine and inorganic phosphate by snake venom nucleotidase; the liberated phosphate is presumably trapped by lead at or near sites of activity. These observations have been criticized by Breckenridge and Johnson (1969), who point out that the localizations observed in retina, cerebral cortex, and cerebellum are not consistent with the biochemical findings. The use of unfixed tissues and the fact that snake venom contains phosphodiesterase activity also raise serious doubts concerning the validity of these localizations.

A method based on the same principle has recently been used to localize cyclic phosphodiesterase activity with electron microscopy in rat cerebral cortex (Florendo *et al.*, 1971). Apparently, phosphodiesterase activity survives glutaraldehyde fixation, and can be demonstrated in postsynaptic nerve endings. Inhibition by theophylline, as well as the use of several other controls, suggests that this method localizes the actual sites of phosphodiesterase activity in brain cells. The details of this procedure are given in the above reference.

A similar method has apparently been used to localize cyclic phosphodiesterase activities in bacteria (Wetzel *et al.*, 1970). However, the accompanying electron micrographs are difficult to evaluate, since they show marked diffusion of the reaction product at the bacterial surface.

It is interesting to note that a lead method for the cytochemical demonstration of adenyl cyclase has also been developed recently (Reik *et al.*, 1970).

"CARBAMYL PHOSPHATASE"

The localization of a phosphatase activity hydrolyzing carbamyl phosphate, an intermediate in the formation of urea in the liver, has been reported by Mizutani and Fujita (1968). The reaction product is found in the endoplasmic reticulum, nuclear envelope, and lysosomes; the lysosomal localization, however, appears to be due to the presence of acid phosphatase activity. The distribution of carbamyl phosphatase activity is virtually identical to that of hepatic glucose-6-phosphatase activity. Further studies are therefore required in order to determine the relationship between these two activities. Details of the procedure are given in the above reference.

CONCLUDING REMARKS

It is evident from the foregoing discussion that the problems of localizing phosphatase activities at the ultrastructural level fall into two separate yet interrelated categories. First, there are the technical difficulties related primarily to the effects of fixation and to the penetration of reagents into tissues; these problems can be largely circumvented with present techniques by following suitable precautions, as discussed in this chapter. Second—and more important—are those limitations that constitute an inherent part of the metal salt procedures. How much improvement can be obtained in reactions of this type remains to be seen, but certainly any modifications that can reduce extraneous precipitate or yield reaction products of smaller dimensions and more uniform size will serve to increase resolution. The reaction products of azo dye methods appear to possess some of these characteristics, although the methods themselves have had limited success to date. The synthesis of additional diazonium salts that produce azo dyes having appropriate chelation properties can be expected to broaden the usefulness of these procedures.

There is little question that modifications in tissue preparation and other conventional cytochemical procedures currently available will continue to provide important information on the localization of phosphatase activities. However, it seems unlikely that such methods will be capable of achieving the increased resolution and greater specificity that are needed to identify a broader spectrum of phosphatases and their isozymes. New developments in methodology may be required to meet these goals.

One of the most promising of the newer methods utilizes immune reactions to localize cell surface and intracellular antigens. These methods depend on the application of specific antibody conjugated to electron opaque markers such as ferritin or to enzymes such as peroxidase whose activity can be visualized by cytochemical means (for a review, see Avrameas, 1970 and Sternberger, this volume). Although not yet extensively applied to the localization of phosphatases, owing partly to difficulties in obtaining highly purified enzyme preparations, the immune reaction, with its exquisite sensitivity, appears to offer one of the most attractive approaches to the localization and differentiation of the many phosphatases that now lie beyond the reach of current methods.

ACKNOWLEDGMENTS

I am grateful to my colleagues for their assistance in the preparation of this chapter. Drs. Etienne de Harven, Sidney Goldfischer, and Constance Oliver read the manuscript critically, and offered many useful suggestions. I am especially indebted to Dr. M. Earl Balis for helpful discussions on some of the biochemical aspects, and to Dr. Oliver for providing the electron micrographs shown in Fig. 1. I also thank Miss Mary Lou McKay and Miss Patricia Fuji for their technical help, Mr. William Matz for the photographs, and Mrs. Dorothy Saltzer for typing several versions of the manuscript. This work was supported, in part, by United States Public Health Service Grant CA-08748.

REFERENCES

Ahrens, R., and Weissenfels, N. (1969). Nachweis von adenosin-triphosphatase in den Mitochondrien Kultivierter Hühnerherz-myoblasten. *Histochemie* **19**, 248.

Allen, J. M. (1961). The histochemistry of glucose-6-phosphatase in the epididymis of the mouse. *J. Histochem. Cytochem.* **9**, 681.

Anderson, P. J. (1967). Purification and quantitation of glutaraldehyde and its effect on several enzyme activities in skeletal muscle. *J. Histochem. Cytochem.* **11**, 652.

Arborgh, B., Ericsson, J. L. E., and Helminen, H. (1971). Inhibition of renal acid phosphatase and aryl sulfatase activity by glutaraldehyde fixation. *J. Histochem. Cytochem.* **19**, 449.

Ashworth, C. T., Luibel, F. J., and Stewart, S. C. (1963). The fine structural localization of adenosine triphosphatase in the small intestine, kidney, and liver of the rat. *J. Cell Biol.* **17**, 1.

Avrameas, S. (1970). Immunoenzyme techniques: enzymes as markers for the localization of antigens and antibodies. *Int. Rev. Cytol.* **27**, 349.

Bainton, D. F., and Farquhar, M. G. (1968). Differences in enzyme content of azurophil and specific granules of polymorphonuclear leukocytes. II. Cytochemistry and electron microscopy of bone marrow cells. *J. Cell Biol.* **39**, 299.

Barka, T., and Anderson, P. J. (1962). Histochemical methods for acid phosphatase using hexazonium pararosanilin as coupler. *J. Histochem. Cytochem.* **10**, 741.

Barrnett, R. J. (1964). Localization of enzymatic activity at the fine structural level. *J. Roy. Micros. Soc.* **83**, 143.

Behnke, O. (1966). Nonspecific deposition of lead in experiments on fine structural localization of enzymatic activity of rat blood platelets. *J. Histochem. Cytochem.* **14**, 432.

Berg, G. G. (1960). Histochemical demonstration of acid trimetaphosphatase and tetrametaphosphatase. *Histochem. Cytochem.* **8**, 92.

Berg, G. G. (1964). The staining of triphosphatases by the chelate removal method. *J. Histochem. Cytochem.* **12**, 341.

Bernhard, W., and Viron, A. (1971). Improved techniques for the preparation of ultrathin frozen sections. *J. Cell Biol.* **49**, 731.

Brandes, D., Zetterqvist, H., and Sheldon, H. (1956). Histochemical techniques for electron microscopy: alkaline phosphatase. *Nature* **177**, 382.

Breckenridge, B. McL., and Johnson, R. E. (1969). Cyclic 3′,5′-nucleotidase phosphodiesterase in brain. *J. Histochem. Cytochem.* **17**, 505.

Byczkowska-Smyk, W., and Bernhard, W. (1960). Essais de cytochimie ultrastructurale. Recherche de la phosphatase alcaline dans le rein du rat à l'aide du microscope électronique. *Comptes. Rendus. Acad. Sci.* (Paris) **251**, 3085.

Chase, W. H. (1963). The demonstration of alkaline phosphatase in frozen dried mouse gut in the electron microscope. *J. Histochem. Cytochem.* **11**, 96.

Cheung, W. Y. (1970). Cyclic nucleotide phosphodiesterase. *In* Role of Cyclic AMP in Cell Function. *Adv. in Biochem. Psychopharm.* **3**, 51.

Chiquoine, A. D. (1953). The distribution of glucose-6-phosphatase in the liver and kidney of the mouse. *J. Histochem. Cytochem.* **1**, 429.

Chiquoine, A. D. (1955). Further studies on the histochemistry of glucose-6-phosphatase. *J. Histochem. Cytochem.* **3**, 471.

Daems, W. Th. (1962). Mouse liver lysosomes and storage: A morphological and histochemical study. (Thesis, Univ. Leiden, Holland) Drukkery̆ Luctor et Emergo, Leiden.

de Thé, G. (1965). Méthode au plomb pour la mise en evidence de la phosphatase alcaline en microscopie électronique. *J. Microscopie* **4**, 130.

Ericsson, J. L. E. (1966). On the fine structural demonstration of glucose-6-phosphatase. *J. Histochem. Cytochem.* **14**, 361.

Ericsson, J. L. E., and Biberfeld, P. (1967). Studies on aldehyde fixation: Fixation rates and their relation to fine structure and some histochemical reactions in liver. *Lab. Invest.* **17**, 281.

Ericsson, J. L. E., and Trump, B. F. (1965). Observations on the application to electron microscopy of the lead phosphate technique for the demonstration of acid phosphatase. *Histochemie* **4**, 470.

Essner, E., Fogh, J., and Fabrizio, P. (1965). Localization of mitochondrial adenosine triphosphatase activity in cultured human cells. *J. Histochem. Cytochem.* **13**, 647.

Essner, E., and Novikoff, A. B. (1961). Localization of acid phosphatase activity in hepatic lysosomes by means of electron microscopy. *J. Biophys. Biochem. Cytol.* **9**, 773.

Essner, E., Novikoff, A. B., and Masek, B. (1958). Adenosine triphosphatase and 5'-nucleotidase activities in the plasma membrane of liver cells as revealed by electron microscopy. *J. Biophys. Biochem. Cytol.* **4**, 711.

Essner, E., Novikoff, A. B., and Quintana, N. (1965). Nucleoside phosphatase activities in rat cardiac muscle. *J. Cell Biol.* **25**, 201.

Farquhar, M. G., and Palade, G. E. (1965). Cell junctions in amphibian skin. *J. Cell Biol.* **26**, 263.

Farquhar, M. G., and Palade, G. E. (1966). Adenosine-triphosphatase localization in amphibian epidermis. *J. Cell Biol.* **30**, 359.

Fitzsimons, J. T. R., Gibson, D. W., and Barrnett, R. J. (1970). A new one-step method for the fine structural localization of acid phosphatase. *J. Histochem. Cytochem.* **18**, 673.

Florendo, N. T., Barrnett, R. J., and Greengard, P. (1971). Cyclic 3',5'-nucleotide phosphodiesterase: Cytochemical localization in cerebral cortex. *Science* **173**, 745.

Ganote, C. E., Rosenthal, A. S., Moses, H. L., and Tice, L. W. (1969). Lead and phosphate as sources of artifact in nucleoside phosphatase histochemistry. *J. Histochem. Cytochem.* **17**, 641.

Gillis, J. M., and Page, S. G. (1967). Localization of ATPase activity in striated muscle and probable sources of artifact. *J. Cell Sci.* **2**, 113.

Goldfischer, S., Essner, E., and Novikoff, A. B. (1964). The localization of phosphatase activities at the level of ultrastructure. *J. Histochem. Cytochem.* **12**, 72.

Goldfischer, S., Essner, E., and Schiller, B. (1971). Nucleoside diphosphatase and thiamine pyrophosphatase activities in the endoplasmic reticulum and Golgi apparatus. *J. Histochem. Cytochem.* **19**, 349.

Gomori, G. (1939). Microtechnical demonstration of phosphatase in tissue sections. *Proc. Soc. Exp. Biol. Med.* **42**, 23.

Gomori, G. (1952). *Microscopic Histochemistry: Principles and Practice.* University of Chicago Press.

Grossman, I. W., and Heitkamp, D. H. (1968). Electron microscopic localization of mitochondrial adenosine triphosphatase activity. *J. Histochem. Cytochem.* **16**, 645.

Hanker, J. S., Deb, C., Wasserkrug, H. L., and Seligman, A. M. (1966). Staining tissue for light and electron microscopy by bridging metals with multidentate ligands. *Science* **152**, 1631.

Hanker, J. S., Seaman, A. R., Weiss, L. P., Ueno, H., Bergman, R. A., and Seligman, A. M. (1964). Osmiophilic reagents: New cytochemical principle for light and electron microscopy. *Science* **146**, 1039.

Hayat, M. A. (1970). *Principles and Techniques of Electron Microscopy: Biological Applications,* Vol. 1. Van Nostrand Reinhold Company, New York.

Hayat, M. A. (1972). *Basic Electron Microscopy Techniques.* Van Nostrand Reinhold Company, New York.

Holt, S. J. (1959). Factors governing the validity of staining methods for enzymes and their bearing upon the Gomori acid phosphatase technique. *Exp. Cell Res. Supply* **7**, 1.

Holt, S. J., and Hicks, R. M. (1961a). Studies on formalin fixation for electron microscopy and cytochemical staining purposes. *J. Biophys. Biochem. Cytol.* **11**, 31.

Holt, S. J., and Hicks, R. M. (1961b). The localization of acid phosphatase in rat liver cells as revealed by combined cytochemical staining and electron microscopy. *J. Biophys. Biochem. Cytol.* **11**, 47.

Holt, S. J., and Hicks, R. M. (1962). Specific staining methods for enzyme localization at the subcellular level. *Brit. Med. Bull.* **18**, 214.

Holt, S. J., and Hicks, R. M. (1966). The importance of osmiophilia in the production of stable azoindoxyl complexes of high contrast for combined enzyme cytochemistry and electron microscopy. *J. Cell Biol.* **29**, 361.

Hopwood, D. (1967). Some aspects of fixation with glutaraldehyde: A biochemical and histochemical comparison of the effects of formaldehyde and glutaraldehyde fixation on various enzymes and glycogen with a note on penetration of glutaraldehyde into liver. *J. Anat.* **101**, 83.

Hopwood, D. (1969). Fixatives and fixation: a review. *Histochem. J.* **1**, 323.

Hugon, J. S. (1970). Ultrastructural differentiation and enzymatic localization of phosphatases in the developing duodenal epithelium of the mouse. *Histochemie* **22**, 109.

Hugon, J. S., and Borgers, M. (1966). A direct lead method for the electron microscopic visualization of alkaline phosphatase activity. *J. Histochem. Cytochem.* **14**, 429.

Hugon, J. S., Borgers, M., and Maestracci, D. (1970). Glucose-6-phosphatase and thiamine pyrophosphatase activities in the jejunal epithelium of the mouse. *J. Histochem. Cytochem.* **18**, 361.

Hündgen, M. (1970). Beitrag zur klarung der spezifität thiaminpyrophosphat und nukleosiddiphosphate spaltender enzyme. *Histochemie* **22**, 376.

Iglesias, J. R., Bernier, R., and Simard, R. (1971). Ultracryotomy: A routine procedure. *J. Ultrastruct. Res.* **36**, 271.

Jacobsen, N. O., and Jørgensen, P. L. (1969). A quantitative biochemical and histochemical study of the lead method for localization of adenosine triphosphate-hydrolyzing enzymes. *J. Histochem. Cytochem.* **17**, 443.

Janigen, D. T. (1965). The effects of aldehyde fixation on acid phosphatase activity in tissue blocks. *J. Histochem. Cytochem.* **13**, 476.

Kanamura, S. (1971). Demonstration of glucose-6-phosphatase activity in hepatocytes following transparenchymal perfusion fixation with glutaraldehyde. *J. Histochem. Cytochem.* **19**, 386.

Kaplan, S., and Novikoff, A. B. (1959). The localization of ATPase activity in the rat kidney: Electron microscopic examination of reaction product in formol-calcium-fixed frozen sections. *J. Histochem. Cytochem.* **7**, 295.

Karnovsky, M. J. (1965). A formaldehyde-glutaraldehyde fixative of high osmolarity for use in electron microscopy. *J. Cell Biol.* **27**, 137A.

Lazarus, S. S., and Barden, H. (1964a). Ultramicroscopic localization of mitochondrial adenosine triphosphatase. *J. Ultrastruct. Res.* **10**, 189.

Lazarus, S. S., and Barden, H. (1964b). Pancreatic β-cell glucose-6-phosphatase: substrate specificity and submicroscopic distribution. *J. Histochem. Cytochem.* **12**, 792.

Leduc, E. H., Bernhard, W., Holt, S. J., and Tranzer, J. P. (1967). Ultrathin frozen sections. II. Demonstration of enzymic activity. *J. Cell Biol.* **34**, 773.

Leskes, A., Siekevitz, P., and Palade,.G. E. (1971). Differentiation of endoplasmic reticulum in hepatocytes. I. Glucose-6-phosphatase distribution *in situ*. *J. Cell Biol.* **49**, 264.

Liptrap, W. H. (1968). Colored electron micrographs: Production and application to distinguish objects of similar density and structure. *J. Ultrastruct. Res.* **25**, 417.

Livingston, D. C., Coombs, M. M., Franks, L. M., Maggi, V., and Gahan, P. B. (1969). A lead phthalocyanin method for the demonstration of acid hydrolases in plant and animal tissues. *Histochemie* **18**, 48.

Lowenstein, J. M. (1958). Transphosphorylation catalyzed by divalent metal ions. *Biochem. J.* **70**, 222.

Marchesi, V. T., and Palade, G. E. (1967). The localization of Mg-Na-K-activated adenosine triphosphatase activity on red cell ghost membranes. *J. Cell Biol.* **35**, 385.

Mayahara, H., Hirano, H., Saito, T., and Ogawa, K. (1967). The new lead citrate method for the ultracytochemical demonstration of non-specific alkaline phosphatase (orthophosphoric monoester phosphohydrolase). *Histochemie* **11**, 88.

Mietkiewski, K., Domka, F., Malendowicz, L., and Malendowicz, J. (1970). Studies on ATP hydrolysis in medium for histochemical demonstration of ATPase activity. *Histochemie* **24**, 343.

Miller, F., and Palade, G. E. (1964). Lytic activities in renal protein absorption droplets: An electron microscopical cytochemical study. *J. Cell Biol.* **23**, 519.

Mizutani, A. (1966). Modification of lead method for demonstration of alkaline phosphatase activity with light and electron microscopy. *Acta Tuberculosea Japonica* **16**, 34.

Mizutani, A., and Barrnett, R. J. (1965). Fine structural demonstration of phosphatase activity at pH 9.0. *Nature* **206**, 1001.

Mizutani, A., and Fujita, H. (1968). Hydrolysis of carbamyl phosphate in the endoplasmic reticulum and nuclear envelope of rat liver. *J. Histochem. .Cytochem.* **16**, 546.

Mölbert, E., Duspiva, F., and von Deimling, O. H. (1960a). Die histochemische lokalisation der phosphatase in der tubulusepithelzelle der mauseniere im elektronenmikroskopischen bild. *Histochemie* **2**, 5.

Mölbert, E. R. G., Duspiva, F., and von Deimling, O. H. (1960b). The demonstration of alkaline phosphatase in the electron microscope. *J. Biophys. Biochem. Cytol.* **7**, 387.

Molnár, J. (1952). The use of rhodizonate in enzymatic histochemistry. *Stain Technol.* **27**, 221.

Moses, H. L., and Rosenthal, A. S. (1967). On the significance of lead-catalyzed hydrolysis of nucleoside phosphates in histochemical systems. *J. Histochem. Cytochem.* **15**, 354.

Moses, H. L., and Rosenthal, A. S. (1968). Pitfalls in the use of lead ion for histochemical localization of nucleoside phosphatases. *J. Histochem. Cytochem.* **16**, 530.

Moses, H. L., Rosenthal, A. S., Beaver, D. L., and Schuffman, S. S. (1966). Lead ion and phosphatase histochemistry. II. Effect of ATP hydrolysis by lead ion on the histochemical localization of ATPase activity. *J. Histochem. Cytochem.* **14**, 702.

Novikoff, A. B. (1959). The proximal tubule cell in experimental hydronephrosis. *J. Biophys. Biochem. Cytol.* **6**, 136.

Novikoff, A. B. (1963). Lysosomes in the physiology and pathology of cells: Contributions of staining methods. In *Ciba Foundation Symposium on Lysosomes* (de Reuck, A. V. S. and Cameron, M. P. eds.), p. 36. Little, Brown and Co., Boston.

Novikoff, A. B. (1964). Membrane-bound enzymes. *Abstracts, 6th Int. Congr. Biochem.: Cellular Organization*, p. 609.

Novikoff, A. B. (1967a). Lysosomes in nerve cells. In *The Neuron*, p. 319. Elsevier Publishing Co., New York.

Novikoff, A. B. (1967b). Enzyme localizations with Wachstein-Meisel procedures: real or artifact. *J. Histochem. Cytochem.* **15**, 353.

Novikoff, A. B. (1970). Their phosphatase controversy: Love's labours lost. *J. Histochem. Cytochem.* **18**, 916.

Novikoff, A. B., Essner, E., Goldfischer, S., and Heus, M. (1962). Nucleoside phosphatase activities of cytomembranes. *Symp. Intern. Soc. Cell Biol.* **1**, 149.

Novikoff, A. B., and Goldfischer, S. (1961). Nucleoside-phosphatase activity in the Golgi apparatus and its usefulness for cytological studies. *Proc. Nat. Acad. Sci. USA* **47**, 803.

Novikoff, A. B., Quintana, N., Villaverde, H., and Forschirm, R. (1966). Nucleoside phosphatase and cholinesterase activities in dorsal root ganglia and peripheral nerve. *J. Cell Biol.* **29**, 525.

Ogawa, K., and Mayahara, H. (1969). Intramitochondrial localization of adenosine triphosphatase activity. *J. Histochem. Cytochem.* **17**, 487.

Oledzka-Slolwinska, H., Creemers, J., and Desmet, V. (1967). Cytidine monophosphate as substrate for the electron microscopic visualization of alkaline phosphatase activity. *Histochemie* **9**, 320.

Padykula, H. A., and Herman, E. (1955). Factors affecting the activity of adenosine triphosphatase and other phosphatases as measured by histochemical techniques. *J. Histochem. Cytochem.* **3**, 161.

Pearse, A. G. E. (1963). Some aspects of the localization of enzyme activity with the electron microscope. *J. Roy. Micros. Soc.* **81**, 107.

Persijn, J. P., Daems, W. Th., de Man, J. C. H., and Meijer, E. A. F. H. (1961). The demonstration of adenosine triphosphatase activity with the electron microscope. *Histochemie* **2**, 372.

Podolsky, R. J. (1968). Deposit formation in muscle fibers following contraction in the presence of lead. *J. Cell Biol.* **39**, 197.

Pullman, E. M., Penefsky, H. S., Datta, A., and Racker, E. (1960). Partial resolution of the enzymes catalyzing oxidative phosphorylation. I. Purification and properties of soluble dinitrophenol-stimulated adenosine triphosphatase. *J. Biol. Chem.* **235**, 3322.

Reale, E. (1962). Electron microscopic localization of alkaline phosphatase from material prepared with cryostat microtome. *Exp. Cell Res.* **26**, 210.

Reale, E., and Luciano, L. (1964). A probable source of errors in electronhistochemistry. *J. Histochem. Cytochem.* **12**, 713.

Rechardt, L., and Kokko, A. (1967). Electron microscopic observations on the mitochondrial adenosine triphosphatase in the rat spinal cord. *Histochemie* **10**, 278.

Reik, L., Petzold, G. L., Higgins, J. A., Greengard, P., and Barrnett, R. J. (1970). Hormone-sensitive adenyl cyclase: cytochemical localization in rat liver. *Science* **168**, 382.

Rosen, S. I. (1969). The localization of glucose-6-phosphate hydrolyzing enzyme in hepatocytes, red blood cells, and leucocytes in the liver of the newborn rat. *J. Anat.* **105**, 579.

Rosenbaum, R. M., and Rolon, C. I. (1962). Species variability and the substrate specificity of intracellular acid phosphatases: A comparison of the lead-salt and azo-dye methods. *Histochemie* **3**, 1.

Rosenthal, A. S., Moses, H. L., Beaver, D. L., and Schuffman, S. S. (1966). Lead ion and phosphatase histochemistry. I. Nonenzymatic hydrolysis of nucleoside phosphates by lead ion. *J. Histochem. Cytochem.* **14**, 698.

Rosenthal, A. S., Moses, H. L., and Ganote, C. E. (1970). Interpretation of phosphatase cytochemical data. *J. Histochem. Cytochem.* **18**, 915.

Rosenthal, A. S., Moses, H. L., Ganote, C., and Tice, L. (1969a). The participation of nucleotide in the formation of phosphatase reaction product: A chemical and electron microscope autoradiographic study. *J. Histochem. Cytochem.* **17**, 839.

Rosenthal, A. S., Moses, H. L., Tice, L., and Ganote, C. E. (1969b). Lead ion and phosphatase histochemistry. III. The effects of lead and adenosine triphosphate concentration on the incorporation of phosphate into fixed tissue. *J. Histochem. Cytochem.* **17**, 608.

Sabatini, D. D., Bensch, K., and Barrnett, R. J. (1963). Cytochemistry and electron microscopy: The preservation of cellular ultrastructure and enzymatic activity by aldehyde fixation. *J. Cell Biol.* **17**, 19.

Saito, T., and Ogawa, K. (1968). Ultracytochemical demonstration of D-fructose-1,6-diphosphatase (D-fructose-1,6-diphosphate 1-phosphohydrolase) activity in the rat liver using lead citrate as capture reagent. *J. Microscopie* **7**, 521.

Scarpelli, D. G., and Kanczak, N. M. (1965). Ultrastructural cytochemistry: principles, limitations, and applications. *Int. Rev. Exptl. Path.* **4**, 55.

Seeman, P. M., and Palade, G. E. (1967). Acid phosphatase localization in rabbit eosinophils. *J. Cell Biol.* **34**, 745.

Shanta, T. R., Woods, W. D., Waitzman, M. B., and Bourne, G. H. (1966). Histochemical method for localization of cyclic 3′,5′-nucleotide phosphodiesterase. *Histochemie* **7**, 177.

Sheldon, H., Zetterqvist, H., and Brandes, D. (1955). Histochemical reactions for electron microscopy: acid phosphatase. *Exp. Cell Res.* **9**, 592.

Skou, J. C. (1962). Preparation from mammalian brain and kidney of the enzyme system involved in active transport of Na+ and K+. *Biochem. Biophys. Acta* **58**, 314.

Smith, R. E. (1970). Comparative evaluation of two instruments and procedures to cut non-frozen sections. *J. Histochem. Cytochem.* **18**, 590.

Smith, R. E., and Fishman, W. H. (1969). p-(Acetoxymercuric) aniline diazotate: a reagent for visualizing the naphthol AS-BI product of acid hydrolase action at the level of the light and electron microscope. *J. Histochem. Cytochem.* **17**, 1.

Sutherland, E. W., and Rall, T. W. (1960). The relation of adenosine 3′,5′-phosphate, and phosphorylase to the actions of catecholamines and other hormones. *Pharmacol. Rev.* **12**, 265.

Szmigielski, S. (1971). The use of dextran in phosphatase techniques employing lead salts. *J. Histochem. Cytochem.* **19**, 505.

Takamatsu, H. (1939). Histologische und biochemische studien uber die phosphatase (I. Mitteilung). Histochemische untersuchungsmethodik der phosphatase und deren verteilung in verschiedenen organen und geweben. *Trans. Soc. Path. Jap.* **29**, 492.

Tetas, M., and Lowenstein, J. M. (1963). The effect of bivalent metal ions on the hydrolysis of adenosine di- and triphosphate. *Biochemistry* **2**, 350.

Tice, L. W. (1969). Lead-adenosine triphosphatase complexes in adenosine triphosphatase histochemistry. *J. Histochem. Cytochem.* **17**, 85.

Tice, L. W., and Barrnett, R. J. (1962). The fine structural localization of glucose-6-phosphatase in rat liver. *J. Histochem. Cytochem.* **10**, 754.

Tice, L. W., and Barrnett, R. J. (1965). Diazophthalocyanins as reagents for fine structural cytochemistry. *J. Cell Biol.* **25**, 23.

Torack, R. M. (1965). Adenosine triphosphatase activity in rat brain following differential fixation with formaldehyde, glutaraldehyde, and hydroxyadipaldehyde. *J. Histochem. Cytochem.* **13**, 191.

Tormey, J. McD. (1966). Significance of the histochemical demonstration of ATPase in epithelia noted for active transport. *Nature* **210**, 820.

Tranzer, J. P. (1965). Utilisation du citrate de plomb pour la mise en évidence de la phosphatase alcaline au microscopie électronique. *J. Microscopie* **4**, 409.

Vethamany, V. G., and Lazarus, S. S. (1967). Ultrastructural localization of Mg^{++} dependent dinitrophenol-stimulated adenosine triphosphatase in human blood platelets. *J. Histochem. Cytochem.* **15**, 267.

von Deimling, O. H. (1964). Die darstellung phosphatfreiset-zender enzyme mittels schwermetall-simultan methoden. *Histochemie* **4**, 48.

Vorbrodt, A., and Bernhard, W. (1968). Essais de localisation au microscope électronique de l'activité phosphatasique nucléaire dans de coupes a congélation ultrafines. *J. Microscopie* **7**, 195.

Wachstein, M., and Meisel, E. (1956). On the histochemical demonstration of glucose-6-phosphatase. *J. Histochem. Cytochem.* **4**, 592.

Wachstein, M., and Meisel, E. (1957). Histochemistry of hepatic phosphatases at a physiological pH with special reference to the demonstration of bile canaliculi. *Amer. J. Clin. Path.* **27**, 13.

Wachtel, A., Lehrer, G. M., Mautner, W., Davis, B. J., and Ornstein, L. (1959). Formalin fixation for the preservation of both intracellular ultrastructure and enzymatic activity for electron microscopic studies. *J. Histochem. Cytochem.* **7**, 291.

Wetzel, B. K., Spicer, S. S., Dvorak, H. F., and Heppel, L. A. (1970). Cytochemical localization of certain phosphatases in *Escherichia coli. J. Bacteriol.* **104**, 529.

Wetzel, B. K., Spicer, S. S., and Horn, R. G. (1967). Fine structural localization of acid and alkaline phosphatases in cells of rabbit blood and bone marrow. *J. Histochem. Cytochem.* **15**, 311.

Wills, E. J. (1967). Localization for electron microscopy of nucleoside phosphatases in human liver. *J. Histochem. Cytochem.* **15**, 754.

Wilson, P. P. (1969). Electron microscopic demonstration of two types of mitochondria with different affinities for lead. *Histochemical J.* **1**, 405.

Yamazaki, M., and Hayaishi, O. (1968). Allosteric properties of nucleoside diphosphatase and its identity with thiamine pyrophosphatase. *J. Biol. Chem.* **243**, 2934.

3

Glycosidases
β-Glucosidases,
β-Glucuronidase

I. D. BOWEN

Department of Zoology,
University College,
Cardiff, Wales Great Britain

INTRODUCTION

Glycosidases are enzymes which catalyze the hydrolysis of glycosidic linkages. Such enzymes include various glucoamylases, dextranase, α- and β-amylases, α- and β-glucosidases, galactosidases, neuraminidase, heparinase, lysosyme, α-mannosidase, β-glycosaminidases, and β-glucuronidase. Recent data on the intracellular distribution of some of these enzymes can be obtained from Roodyn (1967) and Dingle and Fell (1969). The majority of the histochemical studies have been confined to α- and β-glucosidases, galactosidases, β-glucosaminidases, and β-glucuronidases. The historical and technical developments underlying these studies have been summarized (Pearse, 1968; Burstone, 1962). Successful electron-cytochemical investigation has been limited largely to β-glucuronidases, although some preliminary work is available on N-acetyl-β-glucosaminidase and β-D-glucopyranosidase.

Numerous problems have arisen in the development of electron-cytochemical techniques. The subject has been reviewed by Barrnett and Palade (1958), Holt and Hicks (1962), Pearse (1962; 1966), Barrnett (1964), Hayat (1970; and in this volume), and others. According to Barrnett and Palade, difficulties arise from the differing requirements of histochemists and electron microscopists.

They pointed out that the most important requirement for the electron microscopist is the preservation of fine structure. This is achieved through fixation, usually with an aldehyde followed by osmium tetroxide. Complete fixation in this manner, however, places the histochemist, especially the enzyme cytochemist, at a distinct disadvantage. His primary objective is to prepare tissues for the demonstration of enzymatic activity, and fixation with osmium tetroxide usually results in the inactivation of the majority of the enzymes.

The introduction of new aldehyde fixatives (Sabatini *et al.*, 1962; 1963; 1964) facilitated enzyme preservation. This was an important step toward solving the problem. Nevertheless, it should be stressed that many enzymes, particularly glycosidases, are significantly inhibited by aldehyde fixatives (Janigan, 1964). If tissues are routinely fixed with glutaraldehyde, it should be remembered that only residual glycosidase will be obtained in the final preparation. The development of physical fixation procedures such as freeze-drying and freeze-substitution (Rebhun, 1972) could prove advantageous in this respect. Such methods allow the preservation of both the enzymatic activity and the fine structure of tissues.

Another type of method, which has been attempted by Barrnett (1959), involves intravital incubation. By using this method, both the fine structure and enzyme are preserved simply by keeping the cells alive while performing the enzyme test. This approach, although of limited applicability, is especially useful for enzymes sensitive to chemical fixation, as in the case for β-glucosidases.

Progress in the localization of glycosidases at the subcellular level has been slow, due to the nonapplicability of the Gomori-based methods. All methods designed for high-resolution cytochemical localization of glycosidases are based upon the production of an azo dye. This development has been largely due to the contribution of Barrnett (1959) and Barrnett and Tice (1963), who were among the first to indicate the general utility of azo dye methods at the subcellular level. The introduction of naphthol AS derivatives as substrates by Burstone (1958; 1962) has also acted as an incentive in this context. Naphthol AS compounds form highly insoluble primary reaction products which make possible the use of high resolution postcoupling techniques.

Azo dye methods are based upon the precipitation of an appropriate, enzymatically released, naphthol or substituted naphthol by means of a suitable diazonium salt. Ideally, the azo dye so formed should be insoluble in all the media used for tissue preparation, and should possess sufficient electron density to be readily observable in the electron microscope. The problem has been tackled by the following three major approaches: the production of an amorphous relatively dense azo dye by employing a diazonium salt of large molecular dimensions; the production of an electron dense azo dye by coupling with a metal-based diazonium salt; and the production of a distinctly crystalline azo dye.

METHODS USING AMORPHOUS AZO DYES

With regard to the relative electron density of non-metal-containing reaction products, theoretical support has been provided by Isenberg (1957), Zeitler and Bahr (1957), and Valentine (1958). Zeitler and Bahr (1957) pointed out that contrast is approximately proportional to density (mass/volume). A useful cytochemical reaction product should therefore have a density greater than the average density of sectioned material. Ornstein (1957) has shown that the maximum density of fixed tissue elements is ~ 1.6 gm/cc; Lehrer and Ornstein (1959) produced an azo dye of greater density by coupling l-naphthol with hexazonium pararosaniline. By employing this method, they were able to demonstrate cholinesterase activity at the subcellular level.

Methods involving amorphous nonmetallic azo dyes are based largely upon the use of diazotized pararosaniline as a coupler. Barrnett (1959) reported the use of diazotized diaminoanthroquinone (bismarck brown) and pararosaniline (basic fuchsin) for the localization of phosphatase activity at the subcellular level. Davis (1959) initially indicated the usefulness of hexazonium pararosaniline for electron cytochemistry, and this was further confirmed by Davis and Ornstein (1959). All these studies employed l-naphthyl substrates, the l-naphthol released during incubation being soluble.

An improvement in the basic method was presented by Bowen (1968a), who used insoluble napththol-AS TR in the role of primary reaction product for the demonstration of acid phosphatase. In an independent investigation, based upon the work of Hayashi et al. (1964) and Fishman et al. (1967), Hayashi et al. (1968) applied the pararosaniline approach to the electron-cytochemical localization of a glycosidase. By using naphthol AS-BI glucuronide as a substrate and hexazonium pararosaniline as the "diazo" reagent, they reported success in localizing β-glucuronidase activity in rat liver. Bowen and Evans (1966), in a brief note, showed that β-glucosidase activity could be demonstrated by coupling 6-bromo-2-naphthol, enzymatically liberated from 6-bromo-2-naphthyl-β-D-glucopyranoside, with hexazonium pararosaniline.

Hexazonium Pararosaniline Method for
β-Glucuronidase (Hayashi et al., 1968)

1. Fix ~ 3 mm^3 tissue specimens for 20 hr at 4°C in methanol-free 4% formaldehyde containing 7.5% sucrose buffered with 0.1 M cacodylate at pH 7.4. Wash the specimens with three changes of the buffer containing sucrose for 2 hr at 4°C.

2. Cut 40 μ thick sections on a freezing microtome and keep them in cold 7.5% sucrose.

3. Treat sections with 70, 100, and 70% ethanols concurrently for 1 min at -15°C and rinse in two changes of cold 7.5% sucrose.

4. Prepare the coupler, hexazonium pararosaniline, as described by Hayashi *et al.* (1964) or Lehrer and Ornstein (1959). Make up two stock solutions: 1) containing 1 gm of pararosaniline (C.I.No. 42500) dissolved in 30 ml hot 2 N HCl, cooled and filtered; and 2) containing 1 gm of sodium nitrite dissolved in 30 ml distilled water. Pararosaniline is diazotized directly before use. Equal quantities of these two solutions are mixed in a small test tube. Diazotization is completed in ~ 1 min at room temperature, and results in a clear straw-colored solution.

5. Incubate sections in a medium containing 0.25 mM naphthol AS-BI-β-glucuronide, 0.1 M acetate buffer, and 1.8 mM hexazonium pararosaniline at pH 5.2 for 20 min at 37°C.

6. Rinse in three changes of 7.5% sucrose and fix in 1% osmium tetroxide, buffered at pH 7.2, for 1 hr.

7. Wash and dehydrate in a graded series of ethanol, and then transfer the sections into absolute ethanol-propylene oxide mixture (1:1) for 10 min. Run sections through two changes of pure propylene oxide each of 10 to 15 min duration, and finally place in propylene oxide-epoxy resin mixture (1:1) for 2 to 12 hr.

8. Embed in Epon (Hayat, 1972) and examine without poststaining.

9. As a control, 0.25 mM of glucosaccharo−1:4 lactone can be added to the incubation medium. The usual "no-substrate" control should also be carried out.

Results and Comments.   Hayashi *et al.* (1968) indicated that the method described above gave rise to an amorphous, moderately electron-dense reaction product associated with the dense bodies in the pericanalicular cytoplasm of the parenchymal liver cell. They reported that the azo dye product was localized preferentially at the edge of the dense bodies, and that reaction was more marked in vacuolated dense bodies. No staining of the endoplasmic reticulum was observed, and the authors suggested that this could possibly be due to a partial loss of azo dye during embedding or to insufficient electron density of the product.

Several difficulties arise when methods using hexazonium pararosaniline are applied. The reaction product, being homogeneous, may obscure underlying ultrastructure (Bowen, 1968b). Livingston *et al.* (1969) have indicated that the method provides only sufficient resolution to investigate large subcellular organelles, and have suggested that the final azo dye formed is partly soluble in dehydration and embedding media. This partial solubility does not appear to be a consistent feature; in some tissues, the final product is insoluble (Bowen, 1968a).

The major difficulty with the above method concerns the density of the azo dye. Since interpretation of the results depends upon a comparison of the relative electron density of organelles, careful parallel recording of test and control sections is essential. Consistent photographic treatment is necessary to reach any meaningful conclusions. Even so, it is not possible to guarantee that sections of the same thickness are being compared. Consequently, the method tends to be rather laborious.

Localization of β-Glucosidase

No recent reviews of the localization of β-glucosidase activity are available. By using biochemical techniques, β-glucosidases have been shown to occur in many organisms. They have been studied mainly in plants, almond emulsin being an important source of the enzyme. Steensholt and Veibel (1943) demonstrated a β-glucosidase in the mucosa of the small intestine and duodenum of the pig. The enzyme has been purified from *Stachybotryx atra* (Jermyn, 1955) and *Myrothecium verrucaria* (Hash and King, 1958). Kihara *et al.* (1961) have identified a ribosomal bound β-glucosidase in yeast. This enzyme has been further investigated by Hauge *et al.* (1961).

Dahlqvist and Thomson (1965) isolated a specific rat intestinal β-glucosidase that catalyzed the hydrolysis of 6-bromo-2-naphthyl-β-D-glucopyranoside. Evans (1956) demonstrated β-glucosidase activity in the gut of the blowfly *Calliphora erythrocephala*. Newcomer (1954) identified the enzyme in the digestive juice of the cockroach, *Periplaneta americana*. Evans and Payne (1964) demonstrated β-glucosidase in the gut of the desert locust, *Schistocerca gregaria*. Morgan (1967) has carried out a detailed kinetic study of this enzyme in *Locusta migratoria*.

Relatively few workers have employed histochemical methods for the localization of β-glucosidase activity. Cohen *et al.* (1952) suggested that the preparation of 6-bromo-2-naphthyl-β-D-glucopyranoside would be of use in the development of a histochemical test for this enzyme. Rutenburg *et al.* (1958) successfully employed 6-bromo-2-naphthyl-β-D-galactopyranoside as substrate for the demonstration of β-D-galactosidase activity. Rutenburg *et al.* (1960) also reported that α-glucosidase activity could be demonstrated by a postcoupling procedure with 6-bromo-2-naphthyl-α-D-glucopyranoside as substrate. They recommended tetrazonium diorthoanisidine (fast blue B) as coupler.

Pearson *et al.* (1961) reported success using 3-(5-bromoindolyl)-β-D-glucopyranoside as substrate in an indoxyl type method. Pearse (1960) indicated that the 6-bromo-2-naphthyl glycoside could be used to localize β-glucosidase in fresh tissue sections. The technique was applied to the digestive cells of the migratory locust by James (1971). Recently, Ashford (1970) reported the use of red violet LB and fast garnet GBC salts for coupling β-glucosidase—released bromo-naphthol in plant material; she strongly recommended simultaneous coupling.

In order to improve the resolution of the bromo-naphthol method for β-glucosidase, 6-bromo-2-naphthol liberated on the hydrolysis of glucopyranoside was coupled with hexazonium pararosaniline (Bowen and Evans, 1966). Using this technique in a postcoupling reaction, Bowen (1968c) examined the cytochemical distribution of β-glucosidase produced in locust gut digestive caeca. Owing to the sensitivity of the enzyme to aldehyde fixation, an alternative method was devised involving short intravital incubation of substrate.

Cytochemical Methods (Bowen and Evans, 1966; Bowen, 1968c)

In Vitro Method:

1. Fix 1 cm^3 cubes in 5% phosphate-buffered neutral formalin containing 1% PVP and 7.5% sucrose, or in 3% phosphate-buffered glutaraldehyde for 3 hr at 0 to 4°C.
2. Wash for 12 hr in cold 0.1 M phosphate buffer containing the same percent PVP-sucrose solutions.
3. Quench tissue specimens in liquid air, and cut 10 μ cryostat sections of the tissue fixed as above, or, alternatively, use fresh-frozen tissues.
4. Dissolve the substrate, 6-bromo-2-naphthyl-β-D-glucopyranoside, in 2-methoxy-ethanol (1 mg/ml). Set up reaction vessels containing a final substrate concentration of 0.1 mg/ml in 0.1 M phosphate buffer (pH 5.4). Incubate the sections for up to 3 hr at 37°C and then rinse in the phosphate buffer.
5. Transfer the sections into a coupling medium containing 0.14% hexazonium pararosaniline in 0.1 M phosphate buffer (pH 7.0) at 0°C. Coupling is completed in a few minutes. The hexazonium pararosaniline is prepared as described in the previous method.
6. Rinse the sections in buffer and mount in glycerin jelly.
7. Alternatively, sections rinsed in buffer may be fixed for 1 hr in 1% osmium tetroxide, dehydrated in a series of ethanols, and embedded in Araldite or Epon.
8. Cut 2 μ thick sections of the embedded tissue and examine without a counterstain under the optical microscope.
9. In addition to "no-substrate" and "heat-denatured" controls, some sections should be incubated in solutions of free 6-bromo-2-naphthol (dissolved in methoxyethanol) to ascertain the extent of any selective uptake. Solutions of varying concentrations up to 0.1 mg/ml should be tested in this manner.

Alternative Intravital Method:

The substrate solution may be introduced into any organ where a distinct lumen is available.
1. Anaesthetize the organism.
2. Inject, or introduce by means of a catheter, the incubation medium into the desired lumen. Isolate the lumen by means of ligatures if necessary. Incubate for short periods up to 15 min.
3. After incubation, sacrifice the organism and transfer the tissue into the coupling medium. The production of an intense red azo dye indicates that coupling has been completed.

4. If thin sections are required, cut the tissue into 1 mm³ in ice cold 3% buffered glutaraldehyde. Fix for 1 to 2 hr before washing overnight in the buffer.

5. Fix the tissue specimens again in 1% osmium tetroxide for 1 hr at 0 to 4°C, dehydrate in a series of alcohols and embed in Araldite or Epon.

6. Examine 2 μ thick sections as previously described.

7. Alternatively, if thicker sections are required, fix the tissue for 12 hr in 10% neutral formalin containing 1% PVP and 7.5% sucrose. Wash for an equal period in the buffer, quench in liquid air, and cut 10 μ thick sections in a cryostat.

8. Mount air-dried cryostat sections in glycerin jelly.

Results and Comments. The use of hexazonium pararosaniline as a coupler results in a particulate cytochemical localization of β-glucosidase activity (Fig. 3–1). Objections may, however, still be raised to the method presented here, on the grounds of the unsuitability of the substrate. Defendi (1956) demonstrated specific cellular and tissue affinities for the primary reaction product, 6-bromo-2-naphthol. Ahlqvist (1963; 1964) has indicated that certain naphthols, if present in sufficient concentration, appear to be selectively taken up by lipid material in tissue sections.

If pure substrate is used and adequate control experiments employing a range of free-naphthol concentrations are carried out, a meaningful localization of β-glucosidase activity can be achieved. In locust digestive epithelium, for instance, the enzyme appears to be distributed as small particles of lysosomal dimensions throughout the cell (Fig. 3–1). The particles are sometimes concentrated toward either the base or the apex of the epithelial cell. Other larger particles, possibly secondary lysosomes, can also be seen (Fig. 3–1). Similar results are achieved with the intravital method, although there is some diffuse distribution of the reaction product.

One general objection to the intravital method is the possibility of uptake of free bromo-naphthol from the lumen used as incubation chamber. The short incubation periods, however, should insure that most of the substrate enters the cells as the glycoside. The rapid entry of substrate would also be encouraged by the high initial concentration gradient in favor of its diffusion into the epithelial cells. The penetration of epithelial cells by the substrate solution does not appear to be a problem.

With regard to application at the subcellular level, it would seem advisable to use a substrate which would yield a more insoluble primary reaction product than the bromo-naphthol. Burstone (1962) indicates that the substituted naphthol 6-bromo-2-naphthol has a solubility of 195 μg/ml. This could well lead to diffusion artifacts at the subcellular level. Naphthol AS compounds, on the other hand, have a solubility of less than 1 μg/ml.

Fruitful ultrastructural studies on the localization of β-glucosidase could be carried out if it proved feasible to synthesize an appropriate naphthyl AS-β-

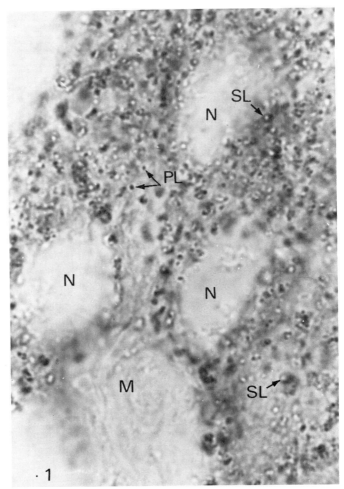

Fig. 3–1. The cytochemical distribution of β-glucosidase in locust gut digestive cells. Enzyme activity, detected by the pararosaniline method, can be seen in small particles (PL), probably primary lyosomes, and also in larger particles (SL), probably representing secondary lysosomes. The nuclei (N) and underlying muscular layers (M) are negative. × 1,500.

glucoside. Rath and Feess (1954) described successful attempts to produce water-soluble naphthol AS compounds in the form of glucosides for dyeing woolen fibers. After application to the fibers, the glucosides could be split and then coupled with diazonium salts to form insoluble azo dyes. Naphthol AS-glucosides can be synthesized by the action of acetobromoglucose on the corresponding naphthol in the presence of quinoline and silver oxide. The tetra-acetate obtained is reesterified to the glucoside and methyl acetate in absolute ethanol with the aid of sodium methylate as a catalyst (Zemplen *et al.*, 1936):

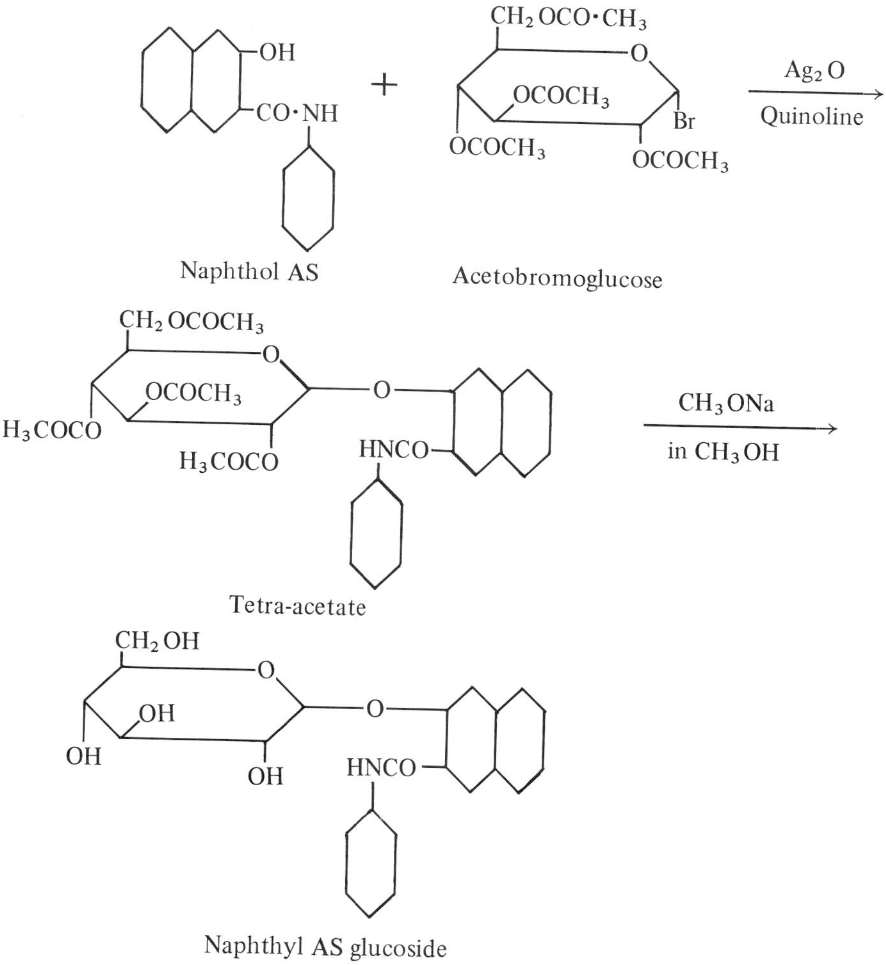

Naphthol AS   Acetobromoglucose

Tetra-acetate

Naphthyl AS glucoside

It would be possible to couple the enzymatically released naphthol with "diazo" compounds such as hexazonium pararosaniline, lead phthalocyanin diazotate, or p-nitrobenzene diazonium-tetrafluoroborate to produce an electron dense azo dye.

METHODS USING METAL-BASED AZO DYES

Burstone (1962) indicated that a successful approach to the ultrastructural localization of enzymatic activities might be afforded by the utilization of substrates which were amenable to metal chelation. Burstone and Weisberger (1961) have produced several azo dyes which could be chelated with metals to form insoluble complexes. Their suggestions have not as yet been tested extensively at the subcellular level.

Tice and Barrnett (1965) described the synthesis of diazophthalocyanins containing chelated metals and employed tri-(4-diazo-) lead phthalocyanin as a capture reagent to localize the sites of acid phosphatase activity in rat kidney tubule cells. They used 1-naphthyl phosphate as substrate in a simultaneous-coupling reaction.

In a constructively critical paper, Livingston *et al.* (1969) gave detailed instructions on the synthesis of lead phthalocyanin, and used the reagent for high-resolution studies of a number of acid hydrolases, including β-glucuronidase. They also examined the solubility of the final reaction product, a naphthol AS-lead phthalocyanin diazotate complex, in dehydration and embedding media. They recommended the use of substituted naphthols as substrates, and stressed the importance of osmication in immobilizing the reaction product. In addition, they recommended Durcupan (Fluka Ltd.) as a dehydration agent.

Smith and Fishman (1968) presented a postcoupling method for the localization of β-glucuronidase activity, utilizing another metal-based diazonium salt, p-(acetoxymercuric) aniline diazotate as coupler. Using naphthyl AS-BI-β-D-glucosiduronic acid as substrate, they investigated the ultrastructural localization of β-glucuronidase in rat liver and preputial gland, and commented on the limitations of the method.

p-(Acetoxymercuric) Aniline Diazotate Method
for β-Glucuronidase (Smith and Fishman, 1968)

1. Fix tissue slices less than 1.5 mm^3 in cold 1.5% singly distilled glutaraldehyde containing 1% sucrose and buffered at pH 7.4 with phosphate or 0.06 M sodium cacodylate.

2. Wash for 24 hr or more at 4°C in 0.1 M cacodylate or veronal acetate buffer (pH 7.4) containing 7% sucrose.

3. Cut 15 μ thick frozen sections or 20 μ thick unfrozen sections (the latter are prepared using a Sorvall TC-2 tissue sectioner), transfer back into a fresh buffer-sucrose wash, and store for a minimum period of 24 hr. Alternatively, 40 μ thick sections can be cut and pretreated with a mixture of acetone, ethanol, and chloroform (6:3:1) for 1 to 2 min before incubation.

4. Prepare a 0.25 mM stock solution of naphthyl AS-BI glucosiduronic acid by dissolving 13.7 mg of the acid in 1 ml 0.05 M NaHCO$_3$ and diluting to 100 ml with 0.1 N acetate buffer (pH 4.5). This stock substrate solution can be

stored for several months at 0 to 4°C, if necessary. The incubation medium is prepared by diluting 16 ml of the stock solution with 0.1 N acetate buffer to give a final volume of 50 ml. Alternatively, an incubation medium of any desired pH can be prepared by appropriate adjustment of the acetate buffer.

5. Dissolve 1 gm p-(acetoxymercuric) aniline (Polysciences or Eastman Organic Chemicals) in 25 ml of 50% acetic acid by shaking in a volumetric flask. Cool the flask by surrounding it with ice at 2 to 4°C. Make up separately a 4% aqueous solution of sodium nitrite by dissolving 2 gm in a 50 ml volumetric flask. Again, hold the flask at 2 to 4°C. After cooling for 45 min, mix the two solutions in equal parts. Stir the solutions for 5 to 20 min in an ice bath to allow diazotization to take place. After diazotization, filter the mixture through a Watman No. 1 paper. Store the diazotized p-(acetoxymercuric) aniline on ice, and use within 12 hr. The coupling solution is prepared by diluting 0.7 ml of the diazotate to 50 ml 1 N phosphate buffer (pH 7.6). The diazotate depresses the pH, which must finally be adjusted to 7.25 with 10 N calcium-free NaOH. This solution must be used within 15 min at temperatures not lower than 15°C.

6. Incubate sections previously prepared from one to several hours (90 min are adequate for rat preputial gland). After incubation, wash the sections for up to 15 min in the buffer wash previously described.

7. Transfer sections into the coupling medium for 2 to 3 min at room temperature (25°C). Wash again in three changes of buffer, each of 5 min duration.

8. Treat sections for 45 min at room temperature with 1% thiocarbohydrazide containing 2% sucrose. Constantly agitate sections during this period. The thiocarbohydrazide solution is made up in a 0.1 M glycine buffer (pH 7.6) through intermittent shaking at 50 to 60°C. The solution must be stored in an oven at 35 to 40°C for short periods before use.

9. Wash sections in two changes of 0.1 M veronal acetate buffer before fixing in 2% cold aqueous osmium tetroxide (pH 7.6) for 90 min at room temperature. Agitate during fixation.

10. In addition to the usual "no-substrate" and "specific inhibitor" controls, 0.01 ml saccharolactone (concentration 20 mg/ml distilled water) per 0.99 ml substrate solution, the following control categories are recommended:

To reduce electron density of lysosomes attributable to metal inclusions, pretreat the sections before incubation for 1 hr at 37°C with 20% EDTA at pH 7.2. Alternatively, sections may be pretreated with 0.5% tetranitromethane in 0.05 M Tris-maleate buffer (pH 8.0) for 3 min at 25°C.

To reduce nonspecific coupling of free diazotate with protein, pretreat sections in 10^{-5} M iodine in 1 N phosphate buffer (pH 7.1) for 30 min at 4°C.

To reduce electron density imparted by the mercury to insoluble SH groups, pretreat sections for 12 to 16 hr at 25°C in 0.1 M N-ethylmaleimide in phosphate buffer (pH 7.0).

11. Dehydrate in a graded series of alcohols. To avoid using propylene oxide,

take sections through the following mixtures of ethanol and Araldite (excluding DMP):

3:1	100% ethanol:Araldite	30 min
1:1	100% ethanol:Araldite	30 min
1:3	100% ethanol:Araldite	45 min
1:7	100% ethanol:Araldite	12 hr

The use of propylene oxide can also be avoided by using acetone instead of ethanol (Hayat, 1970). Finally, embed in Araldite or Epon containing 1 ml DMP accelerator per 50 ml Araldite-DDSA.

Lead Phthalocyanin Method for
β-Glucuronidase (Livingston et al., 1969)

1. Fix fresh tissue blocks (0.2 cm^3) for 2 to 4 hr at 4°C in 2.5% glutaraldehyde made up in 0.1 M cacodylate or phosphate buffer (pH 7.4). The buffer should also contain 0.25 M sucrose.

2. Rinse in the cold buffer for 18 hr or longer.

3. Blot the washed tissues on filter paper and cut 100 μ thick sections with a tissue chopper.

4. Prepare the incubation medium containing 4×10^{-5} M naphthyl AS-β-D-glucuronic acid in 0.1 M acetate buffer (pH 4.2). Sections are placed in this medium for up to 30 min at 37°C.

5. The diazonium salt of tri(4-amino) lead phthalocyanin is prepared by first forming the lead phthalocyanate as follows: 1.3 gm of phthalodinitrile, 5.6 gm 4-nitrophthalodinitrile, and 2 gm of lead metal powder (100 mesh dust) are heated together in 8 ml 1-methylnaphthalene at 205 to 210°C for 1 hr with regular stirring. After cooling, the crude tri(4-nitro) phthalocyanin so formed is ground in 50 ml of ethanol and the solid filtered. The solid is then successively washed in absolute ethanol, 2 N aqueous sodium hydroxide, water, 2 N hydrochloric acid, and acetone. The washing is carried out in each of these components until the filtrate is colorless. The product is dried in air at 100°C, and reduced with 20 gm 1 N sodium dithionite in 200 ml 1 N sodium hydroxide at 60°C for 2 hr. The solid is again filtered off, boiled in 500 ml of a 20% solution of sodium chloride, refiltered, washed in 10% sodium chloride and, finally, in water, before being air-dried at 100°C. The dried solid is placed in a Soxhlet thimble and extracted with boiling tetrahydrofuran under a stream of nitrogen until the extract is colorless. Evaporation of the solvent yields the tri(4-amino) lead phthalocyanin.

6. Suspend 100 mg of the lead phthalocyanin in 3 ml 20% HCl at 0 to 5°C while stirring. Add 2 ml of 1% sodium nitrite (0.5 ml at a time over 5 min

period). Diazotization is completed 10 min after the last addition. Filter the diazo solution, and adjust the pH to 5.0 with cold 1 N sodium hydroxide. Finally, dilute to 10 ml with 0.1 M acetate buffer at the same pH. The solution is now ready to be used as a coupler.

7. After incubation, immerse the sections for 5 to 10 min in cold 0.1 M acetate buffer (pH 4.8 to 5.4) containing 10 mg/ml of the lead phthalocyanin diazotate (concentration calculated as initial suspension of lead phthalocyanin). If a simultaneous couple is attempted, the tissue must be incubated at room temperature to avoid tissue damage. In this case, the same concentration of diazo compound is introduced into the incubation medium before adding the sections.

8. Rinse sections in buffer at 4°C for 15 to 60 min, with repeated changes of the buffer.

9. Fix the tissue in 1% osmium tetroxide for 1 hr.

10. Dehydrate in Durcupan as described by Stäubli (1963), and embed in Araldite (Glauert, 1965) according to the following procedure:

Durcupan:water (1:1)	15-30 min
Durcupan:water (2:1)	15-30 min
Durcupan:water (9:1)	15-30 min
Pure Durcupan (2 changes)	60 min each
Durcupan:Araldite I (70:30)	60 min
Durcupan:Araldite I (50:30)	60 min
Durcupan:Araldite I (30:70)	12 hr
Pure Araldite I	60 min
Araldite II (3 changes)	30 min each

Embed in Araldite II.

11. Examine ultrathin sections without poststaining.

Results and Comments. The p-(acetoxymercuric) aniline diazotate method produces a red azo dye which has a particulate dense appearance in the electron microscope. The technique has limitations, which have been discussed in detail by the authors. The major difficulty is due to nonspecific staining imparted to membranes caused by coupling of the diazotate with structural protein. The diazonium salt also appears to have some affinity for lysosomal compounds. It seems that HgS and other mercurial complexes tend to form in the tissue. This makes interpretation of results more difficult.

Smith and Fishman (1968) have suggested several controls which minimize, but do not entirely remove, these difficulties. They have also pointed out that osmication of the tissue before dehydration increases the electron density of the final reaction product.

In general, the product of the lead phthalocyanin diazotate method is very electron-dense; it compares well with that produced by the Gomori method, and is consequently readily observable at low magnifications. The authors indicate that simultaneous coupling produces severe damage to the fine structure if incubation times in excess of 30 min at 37°C are used. This is thought by them to be due to a temperature-dependent decomposition of the diazotate, giving rise to nitrogen gas and insoluble lead hydroxyphthalocyanin.

Since the primary reaction product is in any case largely insoluble, a post-coupling procedure can be strongly recommended. The incorporation of lead into the diazonium salt means that nonspecific staining similar to that obtained by the Gomori lead-salt technique can occur when this method is used. Particular attention must be given to "no-substrate" controls. Indeed, metal-based azo dye methods are on the whole as open to similar criticisms as the Gomori-based methods are.

Smith and Fishman (1968) indicated the occurrence of β-glucuronidase activity in the cisternae of the Golgi, endoplasmic reticulum, and within the nuclear envelope of rat liver cells. Lipid droplets in rat preputial gland were also claimed to exhibit high β-glucuronidase activity. This localization, however, should be carefully scrutinized, in view of the findings of Ahlqvist (1963; 1964). Livingston et al. (1969) reported the demonstration of the enzyme activity in lysosomes, ribosomes, and cisternae of the rough endoplasmic reticulum. The results obtained by both techniques support the concept of dual localization of β-glucuronidase in lysosomes and in endoplasmic reticulum, as proposed by Fishman et al. (1967).

METHODS USING CRYSTALLINE AZO DYES

Recently, Bowen and Lloyd (1971) described a technique for the electron-cytochemical localization of the site of carbaryl (1-naphthyl N-methyl carbamate) metabolism. The immediate application of this method was confined to the demonstration of sites of metabolism of the pesticide, 1-naphthyl N-methyl carbamate, which is currently in use for insect and mollusk control. The method is dependent upon the hydrolytic release of 1-naphthol from the pesticide and its subsequent coupling with p-nitrobenzene diazonium-tetrafluoroborate to produce a red azo dye. This dye was found to have a distinctive appearance, and could be easily recognized in the electron microscope. The dye was also found to be insoluble in ethanol, propylene oxide, and Araldite.

It is interesting to note that whereas the above method was not primarily designed to demonstrate enzymatic activity, glucuronidase may well be implicated in the metabolism of the pesticide (Knaak et al., 1965), the metabolized naphthol being released as the glucuronide. The compounds employed in the test and the reactions involved are illustrated below.

Crystalline Azo Dye (coupling probably takes place in either of the positions indicated).

The crystalline appearance of the final azo dye in the electron microscope is shown in Fig. 3–2.

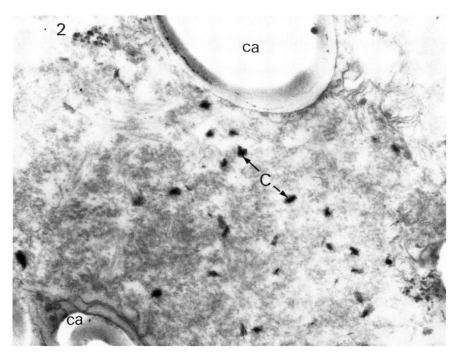

Fig. 3–2. Part of a calcium cell from slug digestive gland. Crystals of azo dye (C) have been formed in the cytoplasm by coupling l-naphthol, metabolically released from carbaryl, with p-nitrobenzene diazonium-tetrafluoroborate. The tissue, of necessity, has been fixed after incubation with pesticide carbaryl and the preservation of fine structure is, therefore, poor. Calcium granules (ca) can, however, be identified. × 20,000.

Bowen (1971) applied the same principle to the localization of acid hydrolases at the subcellular level. The enzymatically released primary reaction product, an insoluble naphthol AS compound, was coupled with p-nitrobenzene diazonium-tetrafluoroborate to produce a dense particulate azo dye. Simultaneous and postcoupling procedures were employed, and a lysosomal and cytoplasmic localization of β-glucuronidase activity (as well as that of acid phosphatase) in rat and slug liver was described. The reactions involved in this test are as follows:

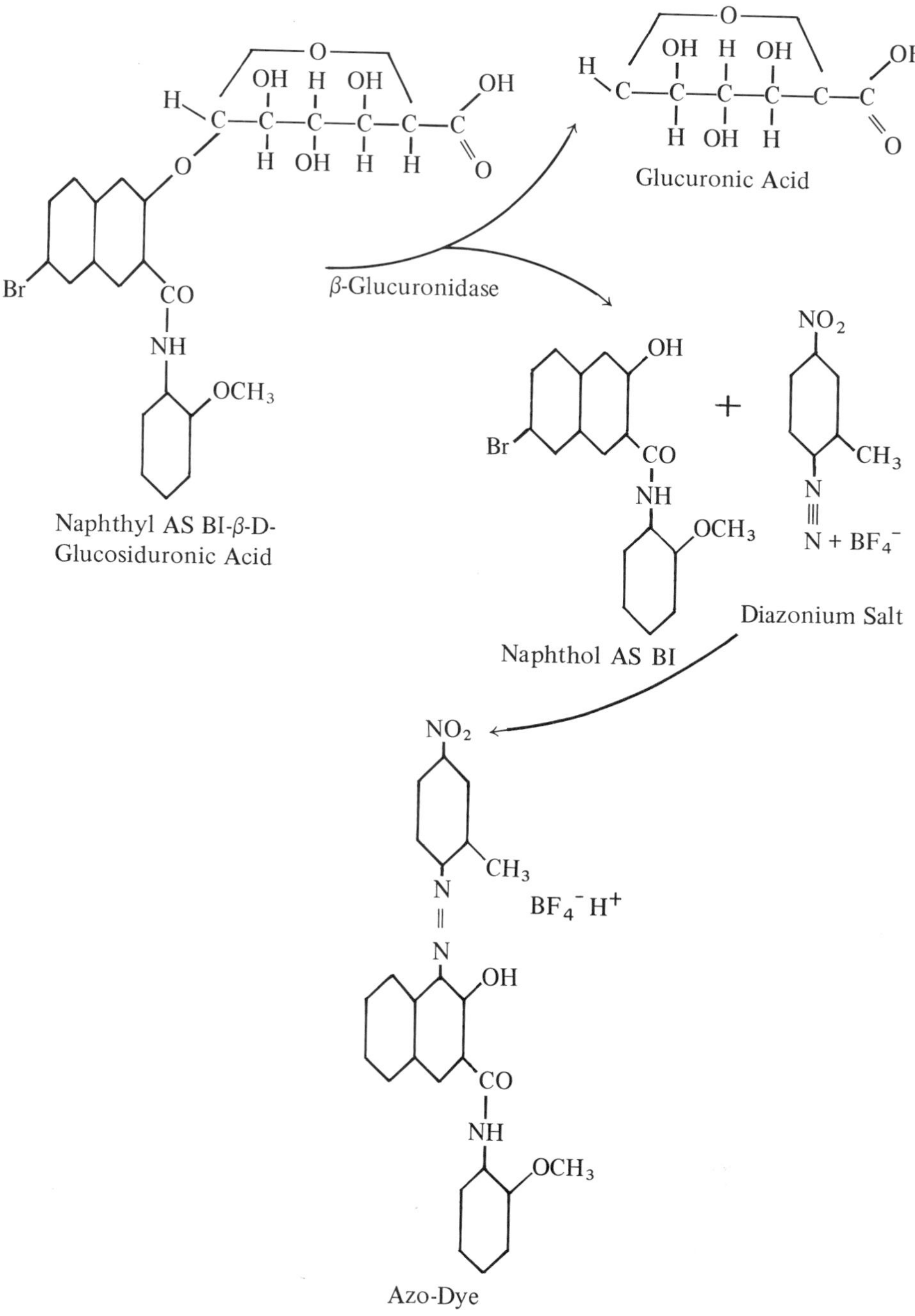

Glucuronic Acid

β-Glucuronidase

Naphthyl AS BI-β-D-
Glucosiduronic Acid

Naphthol AS BI

Diazonium Salt

Azo-Dye

The crystal size produced when the diazonium salt was coupled with a naphthol AS compound (Fig. 3–3) was found to be smaller than that produced when the salt is coupled with 1-naphthol (compare Figs. 3–2 and 3–3), thus enhancing the potential resolving power of the test.

p-Nitrobenzene Diazonium-Tetrafluoroborate Method for β-Glucuronidase (Bowen, 1971)

1. Fix 1 mm^3 fresh tissue cubes for 1 hr in 3% glutaraldehyde buffered with 0.1 M phosphate buffer (pH 7.4).

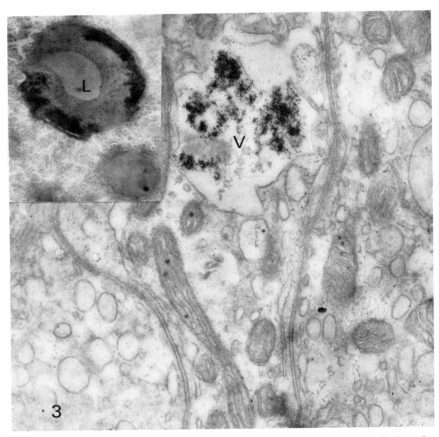

Fig. 3–3. The localization of β-glucuronidase in a slug digestive epithelial cell. Reaction product appears as fine particles of azo dye within the vacuole (V). The azo dye has been produced by coupling naphthol AS BI, enzymatically released from the glucuronide, with p-nitrobenzene diazonium-tetrafluoroborate. × 25,000. Inset: A similar localization produced within a rat liver lysosome (L). × 20,000.

2. After fixing, cut some of the cubes on a Sorvall tissue sectioner to obtain 50 μ thick slices. Wash overnight in the phosphate buffer at 0-4°C.

3. Rinse briefly in 0.1 M acetate buffer (pH 4.5) and transfer tissues into the incubation medium.

4. As recommended by Smith and Fishman (1968), 0.25 mM naphthyl AS BI-glucosiduronic acid (Sigma Ltd.) is used as substrate. Prepare a stock solution by dissolving 13.7 mg of the acid in 1 ml of 0.05 M NaHCO$_3$. Dilute to 100 ml with acetate buffer (pH 4.5) and store in a refrigerator at 0 to 4°C. Prepare working solution by diluting 16 ml of the stock to 50 ml with the same acetate buffer.

5. Incubate the tissue in the substrate solution for periods up to 3 hr at room temperature or 37°C. Transfer the tissue into a coupling medium containing 30 mg of p-nitrobenzene diazonium-tetrafluoroborate (Kodak Ltd.) in 50 ml 0.1 M phosphate buffer (pH 7.0). Coupling takes place over a period of 30 min at room temperature. To carry out a simultaneous couple, add 30 mg of the diazonium salt to every 50 ml of the substrate solution. Since the salt does inhibit enzymatic activity somewhat, postcoupling is recommended.

6. Rinse tissues briefly in the phosphate buffer and fix in 1% osmium tetroxide (pH 7.4) (Millonig, 1961) for 2 to 3 hr at 0 to 4°C.

7. Dehydrate in a series of alcohols and embed in Araldite.

8. Mount ultrathin sections on grids and examine without poststaining.

9. Incubate control batches in substrate-free media using both simultaneous and postcoupling procedures. Incubate some sections in solutions (up to 0.25 mM) of the free naphthol (naphthol AS BI can be dissolved in acetone or dimethyl sulfoxide). The specific inhibitory effect of adding 0.25 mM saccharolactic acid to the incubation medium should also be investigated.

Results and Comments. The final reaction product takes the form of very fine dense particles measuring up to 6 Å across, and can be found associated with the appropriate bearers of enzymatic activity (Figs. 3–3 and 3–4). In the case of rat and slug liver, the reaction product can be found associated with vacuolar elements or lysosomes and also with ribosomes in the cytoplasm.

Although the test is free from the artifacts which normally beset the commonly used lead-salt methods, there are a few difficulties. The first relates to particle size. Since the azo dye formed is so fine, it is sometimes difficult to identify at low magnifications in the electron microscope. It is often necessary to employ magnifications of 16,000 X and above to positively identify the reaction product, particularly if it is sparse. A more serious drawback in certain tissues is that the reaction product overlaps with ferritin-like particles of similar size.

In many tissues, the problem can be satisfactorily resolved by careful comparison of test and control sections. A more conclusive approach is afforded by carrying out parallel biochemical and electron cytochemical analysis, such as

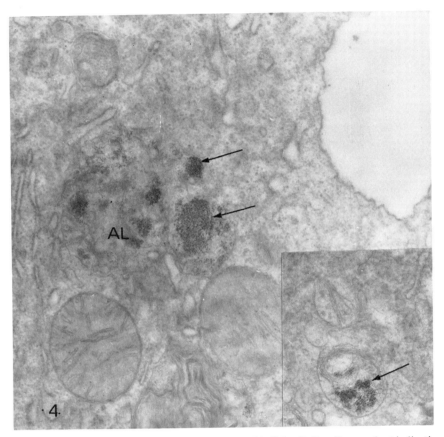

Fig. 3–4. An autolysosome (AL) from a slug gut epithelial cell. Reaction product indicative of β-glucuronidase activity (arrows) can be seen associated with semi-digested membranes within the autolysosome. × 32,000. Inset: An autolysosome from a rat liver cell exhibiting β-glucuronidase activity. × 32,000.

described by Cooper *et al.* (1973) for acid phosphatase. In this study, the same basic reaction has been performed biochemically on identified fractions and also electron cytochemically on intact cells (Figs. 3–5 and 3–6). The organism used in this study was the colorless alga, *Polytomella caeca.*

One of the main advantages of the p-nitrobenzene diazonium-tetrafluoroborate technique is that it is readily applicable to biochemical studies. Its use in colorimetry was established by Miskus *et al.* (1959). Other workers have demonstrated that 1-naphthol may be coupled with p-nitrobenzene diazonium-tetrafluoroborate, and the colored product has been used as a chromatographic indicator and in colorimetry (Zweig and Archer, 1958; Krishna *et al.*, 1962; Johnson, 1963; Finocchiaro and Benson, 1967). The technique thus represents a

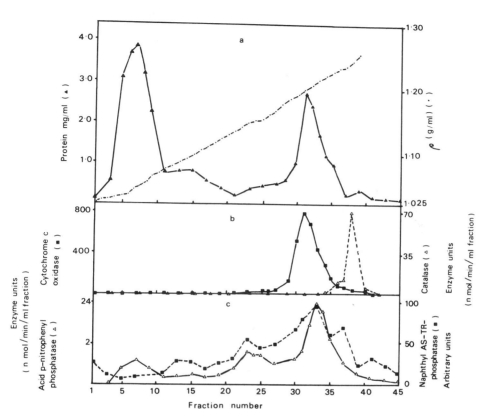

Fig. 3–5. Fractionation of a whole homogenate of *Polytomella caeca* by high-speed zonal centrifugation. The homogenate (20 ml) contained 544 mg total protein. Whole homogenate was diluted 1 in 20, and volumes of diluted homogenate and of fractions taken for assays were as follows: cytochrome c oxidase and catalase, 0.1 M; protein and acid p-nitrophenyl phosphatase, 0.2 ml; naphthyl AS TR phosphatase, 0.5 ml. With regard to the latter, the released naphthol was coupled with p-nitrobenzene diazonium-tetrafluoroborate and the extinction measured at 550 mμ in a Beckman DB spectrophotometer. Centrifugation was accomplished at 35,000 rev/min for 165 min in a BX IV zonal rotor. Specific activities of enzymes in the whole homogenate and recoveries (in parentheses) were as follows: cytochrome c oxidase, 5.6 (60%); catalse, 14.7 (25%), acid phosphatase 7.7 (51%). (*Courtesy R. Cooper and D. Lloyd*)

tool which is potentially applicable in both the fields of biochemistry and cytochemistry.

Using a biochemical and cytochemical approach simultaneously, the technique could be of some value in studying changes in the patterns of enzyme activity. The test works well on plant material (Fig. 3–6). Since the final azo dye is colored, the same reaction can be used at the level of the optical microscope. The technique is particularly useful for electron-cytochemical studies on au-

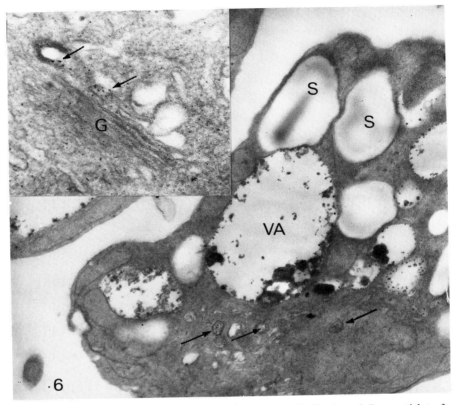

Fig. 3–6. An electron micrograph showing the vacuolar distribution of fine particles of reaction product (arrows) in *Polytomella caeca.* The distribution of this particulate azo dye, produced by coupling enzymatically released naphthol AS TR with p-nitrobenzene diazonium-tetrafluoroborate, reflects the activity of naphthyl AS TR phosphatase within the organism (compare with Fig. 3–5). The starch granules (S) and the large vacuoles (VA) containing lipid droplets are negative. X 30,000. Inset: The Golgi (G) of *Polytomella caeca* showing a slight positive reaction (arrows) for naphthyl AS TR phosphatase. X 200,000. *(Courtesy R. Cooper and D. Lloyd)*

tophagy, since under optimal incubation conditions the reaction product does not completely obscure the underlying fine structure. This means that the investigator has little difficulty when attempting to classify lysosomes, even when they exhibit positive staining (Fig. 3–4).

CONCLUDING REMARKS

To date, very few glycosidases have been studied at the subcellular level, and the majority of the enzymes within the group still await investigation, although a start has been made with β-glucuronidases and β-glucosaminidases. The inhibitory effect of fixation, particularly on glucosidases, remains the major problem to be overcome. For a detailed discussion on the methodology for the localization of β-glucosaminidases, the reader is referred to Pugh (in this volume).

Every electron cytochemical method for the localization of enzymatic activity has its difficulties. This appears to be equally true of azo dye and lead salt techniques. An investigator in this field would be wise to attempt as many methods as possible. It would be profitable to perform parallel analytical biochemical studies, if possible. In this way, cytochemists could contribute much to cell fractionation studies. Used on its own, the electron microscope can be a very limiting instrument. The magnifications involved often lead to an overselective interpretation of data, and can easily give rise to unrepresentative results.

The careful and extensive study of a full array of controls cannot be overemphasized in this context. It is also important to bear in mind that a satisfactory method for one tissue or cell type may not be equally applicable to other tissues or cell types. This question is of particular relevance at the subcellular level, where so many factors can critically affect the localization produced. Indeed, Holt and Hicks (1962) have argued that attempts to localize an enzyme within the subcellular components of different tissues should be regarded as separate problems.

REFERENCES

Ahlqvist, J. (1963). The affinity of some naphthols for tissues before and after extraction of lipids. *Acta Path. Micro. Scand.* **59**, 171.

Ahlqvist, J. (1964). Extraction of some histochemical enzyme substrates from water with ethyl ether. *Ann. Med. Exp. Fenn.* **42**, 85.

Ashford, A. E. (1970). Histochemical localization of β-glycosidases in roots of *Zea mays*. I. A simultaneous coupling azo dye technique for the localization of β-glucosidase and β-galactosidase. *Protoplasma* **71**, 389.

Barrnett, R. J. (1959). The demonstration with the electron microscope of the end products of histochemical reactions in relationship to the fine structure of cells. *Exp. Cell Res. Suppl.* **7**, 65.

Barrnett, R. J. (1964). Localization of enzymatic activity at the fine structural level. *J. Roy. Micros. Soc.* **83**, 143.

Barrnett, R. J., and Palade, G. E. (1958). Applications of histochemistry to electron microscopy. *J. Histochem. Cytochem.* **6**, 1.

Barrnett, R. J., and Tice, L. W. (1963). The Use of the Electron Microscope as a Tool for the Investigation of Problems in Cytochemistry. In *Histochemistry and Cytochemistry* (Wegmann, R. ed.), p. 139. Pergamon Press, New York.

Bowen, I. D. (1968a). Electron cytochemical localization of acid phosphatase activity in the digestive caeca of the desert locust. *J. Roy. Micros. Soc.* **88**, 279.

Bowen, I. D. (1968b). Electron cytochemical studies on autophagy in the gut epithlial cells of the locust, *Schistocerca gregaria. Histochem. J.* **1**, 141.

Bowen, I. D. (1968c). The histochemistry and ultrastructure of the locust digestive caeca. Ph.D. Thesis, University of Wales.

Bowen, I. D. (1971). A high resolution technique for the fine structural localization of acid hydrolases. *J. Microscopy* **94**, 25.

Bowen, I. D., and Evans, W. A. L. (1966). Localization of β-glucosidases in the digestive caeca of the locust, *Schistocerca gregaria*. *Proc. Roy. Micros. Soc.* **1**, 215.

Bowen, I. D., and Lloyd, D. C. (1971). A technique for the electron cytochemical localization of the site of carbaryl metabolism. *J. Invert. Path.* **18**, 183.

Burstone, M. S. (1958). The relationship between fixation and techniques for the histochemical localization of hydrolytic enzymes. *J. Histochem. Cytochem.* **6**, 323.

Burstone, M. S. (1962). *Enzyme Histochemistry and Its Application to the Study of Neoplasms.* Academic Press, New York and London.

Burstone, M. S., and Weisburger, J. (1961). New diazonium components as coupling agents in the demonstration of phosphatases. *J. Histochem. Cytochem.* **9**, 301.

Cohen, R. B., Tsou, K. C., Rutenburg, S. H., and Seligman, A. M. (1952). The colorimetric estimation and histochemical demonstration of β-D-galactosidase. *Biol. Chem.* **195**, 239.

Cooper, R., Bowen, I. D., and Lloyd, D. (1973). Properties and electron microscopical localization of acid phosphatases in *Polytomella caeca* (in preparation).

Dahlqvist, B. B., and Thomson, D. L. (1965). Rat intestinal 6-bromo-2-naphthyl glycosidase and disaccharidase activities. *Arch. Biochem. Biophys.* **109**, 159.

Davis, B. J. (1959). Histochemical demonstration of erythrocyte esterase. *Proc. Soc. Exp. Biol. Med.* **101**, 90.

Davis, B. J., and Ornstein, L. E. (1959). High resolution enzyme localization with a new diazo reagent, hexazonium pararosaniline. *J. Histochem. Cytochem.* **7**, 297.

Defendi, V. (1956). Observations on naphthol staining and the histochemical localization of enzymes by naphthol-azo dye technique. *J. Histochem. Cytochem.* **5**, 1.

Dingle, J. T., and Fell, H. B. (1969). *Lysosomes in Biology and Pathology,* Vols. 1 and 2. North-Holland Publishing Company, Amsterdam and London.

Evans, W. A. L. (1956). Studies on the digestive enzymes of the blowfly, *Calliphora erythrocephala*. I. Carbohydrases. *Expl. Parasit.* **53**, 219.

Evans, W. A. L., and Payne, D. W. (1964). Carbohydrases of the alimentary tract of the desert locust, *Schistocerca gregaria* Forsk. *J. Insect Physiol.* **10**, 657.

Finocchiaro, J. M., and Benson, W. R. (1967). Thin-layer chromatography of some carbamate and phenylurea pesticides. *J. Assoc. Offic. Agri. Chem.* **50**, 888.

Fishman, W. H., Goldman, S. S., and De Lellis, R. (1967). Dual localization of β-glucuronidase in endoplasmic reticulum and in lysosomes. *Nature* **213**, 457.

Glauert, A. M. (1965). The Fixation and Embedding of Biological Specimens. In *Techniques for Electron Microscopy* (Kay, D. H., ed.). Blackwell Science Publications, Oxford.

Hash, J. H., and King, K. W. (1958). Some properties of an aryl-β-glucosidase from culture filtrates of *Myrothecium verrucaria*. *J. Biol. Chem.* **232**, 395.

Hauge, J. G., Macquillan, A. M., Cline, A. L., and Halvorson, H. O. (1961). The effect of glucose repression on the level of ribosomal bound β-glucosidase. *Biochem. Biophys. Res. Commun.* **5**, 267.

Hayashi, M., Nakajima, Y., and Fishman, W. H. (1964). The cytologic demonstration of β-glucuronidase employing naphthol AS-BI glucuronide and hexazonium pararosaniline: a preliminary report. *J. Histochem. Cytochem.* **12**, 293.

Hayashi, M., Shirahama, T., and Cohen, A. S. (1968). Combined cytochemical and electron microscopic demonstration of β-glucuronidase activity in rat liver with the use of a simultaneous coupling azo dye technique. *J. Cell Biol.* **36**, 287.

Hayat, M. A. (1970). *Principles and Techniques of Electron Microscopy: Biological Applications,* Vol. 1. Van Nostrand Reinhold Company, New York.

Hayat, M. A. (1972). *Basic Electron Microscopy Techniques.* Van Nostrand Reinhold Company, New York.

Holt, S. J., and Hicks, R. M. (1962). Combination of Cytochemical Staining Methods for Enzyme Localization with Electron Microscopy. In *The Interpretation of Ultrastructure.* Int. Soc. Cell Biol., (Harris, R. J. C., ed.), p. 193. Academic Press, London and New York.

Isenberg, I. (1957). The use of organic dyes in electron microscopy. *Bull. Math. Biophys.* **19**, 279.

James, E. M. V. (1971). Studies on the histology and histochemistry of the alimentary canal of the migratory locust. MSc., Thesis, University of Wales.

Janigan, D. T. (1964). The effects of aldehyde fixation on β-glucuronidase, β-galactosidase, N-acetyl-β-glucosaminidase, and β-glucosidase in tissue blocks. *Lab. Invest.* **13**, 1038.

Jermyn, M. A. (1955). Enzymatic properties of *Stachybotryx atra* β-glucosidase. *Aust. J. Biol. Sci.* **8**, 541.

Johnson, D. P. (1963). Determination of sevin insecticide residues in fruits and vegetables. *J. Assoc. Offic. Agri. Chem.* **46**, 234.

Kihara, H. K., Hu, A. S. B., and Halvorson, H. O. (1961). The identification of a ribosomal-bound β-glucosidase. *Proc. Nat. Acad. Sci.* **47**, 489.

Krishna, J. G., Dorough, H. W., and Casida, J. E. (1962). Synthesis of N-methyl-carbamates via methyl isocynate C^{14} and chromatographic purification. *J. Agri. Food Chem.* **10**, 462.

Knaak, J. B., Tallant, M. J., Bartley, W. J., and Sullivan, L. J. (1965). The metabolism of carbaryl in the rat, guinea pig, and man. *J. Agri. Food Chem.* **13**, 537.

Lehrer, G. M., and Ornstein, J. E. (1959). A diazo coupling method for the electron microscopic localization of cholinesterase. *J. Biophys. Biochem. Cytol.* **6**, 399.

Livingston, D. C., Cooms, M. M., Franks, L. M., Maggi, V., and Gahan, P. B. (1969). A lead phthalocyanin method for the demonstration of acid hydrolases in plant and animal tissues. *Histochemie* **18**, 48.

Millonig, G. (1961). Advantages of a phosphate buffer for osmium tetroxide solutions in fixation. *J. Appl. Phys.* **32**, 1637.

Miskus, R., Gordon, H. T., and George, D. A. (1959). Colorimeteric determination of 1-naphthyl N-methylcarbamate in agricultural crops. *J. Agri. Food Chem.* **7**, 613.

Morgan, M. R. J. (1967). Kinetic studies on the gut β-glucosidase of *Locusta migratoria migratorioides* (R and R). Ph.D. Thesis, University of Wales.

Newcomer, W. S. (1954). The occurrence of β-glucosidase in the digestive juice of the cockroach, *Periplaneta americana. J. Cell Comp. Physiol.* **43**, 79.

Ornstein, L. E. (1957). "Osmiophilia," fact or fiction? *J. Biophys. Biochem. Cytol.* **3**, 809.

Pearse, A. G. E. (1960). Histochemistry, Theoretical and Applied. 2nd ed. J. and A. Churchill Ltd., London.

Pearse, A. G. E. (1962). Some aspects of the localization of enzyme activity within the electron microscope. *J. Roy. Micros. Soc.* **81**, 107.

Pearse, A. G. E. (1966). The present and future of electron cytochemistry. *J. Histochem. Cytochem.* **14**, 768.

Pearse, A. G. E. (1968). *Histochemistry, Theoretical and Applied,* 3d ed. J. and A. Churchill Ltd., London.

Pearson, B., Andrews, M., and Grose, F. (1961). Histochemical demonstration of mammalian glucosidase by means of 3-(5-bromoindolyl)-β-D-glucopyranoside. *Proc. Soc. Exp. Biol. Med.* **108**, 619.

Pugh, D., and Walker, P. G. (1961). Histochemical localization of β-glucuronidase and N-acetyl-β-glucosamine. *J. Histochem. Cytochem.* **9**, 105.

Rath, H., and Feess, E. (1954). Versuche zur herstellung wasserloslicher naphtole fur das fabren von wolle. *Melliand Text Ber.* **35**, 267.

Rebhun, L. I. (1972). Freeze-Substitution and Freeze-Drying. In *Principles and Techniques of Electron Microscopy: Biological Applications,* Vol. 2 (Hayat, M. A., ed.). Van Nostrand Reinhold Company, New York.

Roodyn, D. B. (1967). *Enzyme Cytology.* Academic Press, London and New York.

Rutenburg, A. M., Goldbarg, J. A., Rutenburg, S. H., and Xang, R. T. (1960). The histochemical demonstration of α-D-glucosidase in mammalian tissues. *J. Histochem. Cytochem.* **8**, 268.

Rutenburg, A. M., Rutenburg, S. H., Monis, B., Teaque, A., and Seligman, A. M. (1958). Histochemical demonstration of β-D-galactosidase in the rat. *J. Histochem. Cytochem.* **6**, 122.

Sabatini, D. D., Bensch, K., and Barrnett, R. J. (1962). New means of fixation for electron microscopy and histochemistry. *Anat. Rec.* **142**, 247.

Sabatini, D. D., Bensch, K., and Barrnett, R. J. (1963). Cytochemistry and electron microscopy: The preservation of cellular ultrastructure and enzymatic activity by aldehyde fixation. *J. Cell Biol.* **17**, 19.

Sabatini, D. D., Miller, F., and Barrnett, R. J. (1964). Aldehyde fixation for morphological and enzyme histochemical studies with the electron microscope. *J. Histochem. Cytochem.* **12**, 57.

Smith, R. E., and Fishman, W. H. (1968). p-(Acetoxymercuric) aniline diazotate: a reagent for visualizing the naphthol AS-BI product of acid hydrolase action at the level of the light and electron microscope. *J. Histochem. Cytochem.* **17**, 1.

Stäubli, W. A. (1963). A new embedding technique for electron microscopy: combining a water soluble epoxy resin (Durcupan) with water insoluble Araldite. *J. Cell Biol.* **16**, 197.

Steensholt, G., and Veibel, S. (1943). On the glucosidases of the intestine of the pig. *Acta Physiol. Scand.* **6**, 62.

Tice, L. W., and Barrnett, R. J. (1965). Diazophthalocyanins as reagents for fine structural cytochemistry. *J. Cell Biol.* **25**, 23.

Valentine, R. C. (1958). Contrast in the electron microscope. *Nature* **181**, 832.

Zeitler, E., and Bahr, G. F. (1957). Contributions to the quantitative interpretation of electron microscope pictures. *Expl. Cell Res.* **12**, 44.

Zemplen, G., Gerecs, A., and Hadacsy, I. (1936). Uber die verseifung acetylierter Kohlenhydrate. *Ber.* **69**, 1827.

Zweig, G., and Archer, T. E. (1958). Residue determination of sevin (1-naphthyl N-methyl carbamate) in wine by cholinesterase inhibition and paper chromatography. *J. Agri. Food Chem.* **6**, 910.

4

Glycosidases N-Acetyl-β-Glucosaminidase

D. PUGH

Department of Zoology,
University of Southhampton,
Southhampton, Great Britain

INTRODUCTION

N-acetyl-β-glucosaminidase is widely distributed in mammalian tissues (Watanabe, 1936; Conchie *et al.*, 1959). The enzyme also occurs in plant tissues (Helfrich and Iloff, 1933), bacteria (Humphery, 1946), mollusks (Neuberger and Pitt Rivers, 1939; Howe and Kabat, 1953; Pugh, 1963), and vertebrates (Wachtler and Pearse, 1966; Pugh, 1967). N-acetyl-β-glucosaminidase is involved in the metabolism of mucosubstances. Oligosaccharides are formed from hyaluronic acid by the action of hyaluronidase, and these are further degraded by β-glucuronidase and N-acetyl-β-glucosaminidase (Linker *et al.*, 1955). N-acetyl-β-glucosaminidase occurs in high concentrations in tissues (e.g., salivary glands, synovial membrane, and respiratory and intestinal epithelia) where a high rate of mucoid metabolism is found.

The intracellular distribution of N-acetyl-β-glucosaminidase has been investigated by biochemical and histochemical methods. Sellinger *et al.* (1960) showed that the enzyme has a lysosomal distribution in isotonic homogenates of rat liver. The subcellular distribution of N-acetyl-β-glucosaminidase in rat kidney was studied by Price and Dance (1967). They found that the enzyme had a bimodal distribution; the greater part of the activity, however, was lysosomal.

Robinson and Stirling (1968) found two distinct forms of N-acetyl-β-glucosaminidase in human spleen, one an acidic, and the other a basic protein; these were named "A" and "B," respectively. The two components showed similar characteristics when assayed with synthetic substrates, and similar reactions to inhibitors. These enzyme forms differed in their stability to heat and changes in pH; the A form was the less stable of the two. Both enzyme forms were found in the lysosomal fraction; the B component was restricted to this fraction, whereas the A component was also present in the supernatant. By isolating these two components from peritoneal macrophages, Robinson and Stirling (1968) demonstrated that both forms of N-acetyl-β-glucosaminidase can occur in a single cell. These two forms corresponded to the forms obtained from homogenates of whole spleen.

The methods available for the histochemical demonstration of N-acetyl-β-glucosaminidase are simultaneous coupling azo dye methods (Pugh and Walker, 1961a and b; Hayashi, 1965) and techniques employing an indoxyl substrate. 5-bromoindoxyl N-acetyl-β-glucosaminide can be used in an oxidation, coupling or metal precipitation technique (Pugh, 1972a). Satisfactory enzyme localization can be obtained using any of these methods. The activity is always found in discrete particles, and the distribution in the tissues is similar but not identical to that of other lysosomal enzymes (Pugh and Walker, 1961a; Pugh, 1964; Hayashi, 1967).

The histochemical results cannot be directly compared with those obtained from fractionated tissue homogenates. Different modes of tissue processing are necessary to preserve biochemical activity and cell morphology, and different enzyme substrates are employed in these two types of studies. Detailed investigations have been made of only a few organs. The bimodal distribution of enzyme activity found in rat kidney may also occur in other tissues, while the activity demonstrated by cytochemical techniques is restricted to subcellular organelles.

One possible explanation of the above results is that the B form of the enzyme activity is the component demonstrated in tissue sections, and all of form A (which is more sensitive to pH changes and to heat) is destroyed during tissue processing. However, it is equally possible that both enzyme forms contribute to the histochemical staining, and the nonparticulate fraction of the A component (if a soluble fraction does exist in the intact cell) is destroyed by fixation because it is more accessible to the fixation fluid.

Fixation cannot be avoided, because indoxyl methods are not generally applicable to fresh tissues (Holt, 1958). Also, fixation is necessary because the preservation of cell structure and the prevention of enzyme diffusion are as important as the preservation of enzyme activity for the definition of active cell sites. The need to use fixed tissues does not impede the study of N-acetyl-β-glucosaminidase, because the enzyme is relatively resistant to fixation; more than 70% of the activity of fresh rat liver survives fixation in formalin calcium chloride (Pugh and Walker, 1961b).

DEVELOPMENT OF N-ACETYL-β-GLUCOSAMINIDASE METHODS

A histochemical technique for the localization of N-acetyl-β-glucosaminidase activity was introduced by Pugh and Walker (1961b), in which $\propto$-naphthyl N-acetyl-β-glucosaminide (Leabeck and Walker, 1957) was used in a simultaneous-coupling azo dye method. The optimum conditions for enzyme preservation and incubation were determined during the development of this method, and have been used in the later techniques without major changes. This method does not yield precise enzyme localization; however, the localization can be improved by using Naphthol AS substrates. Naphthyl AS-LC N-acetyl-β-glucosaminide was employed in a simultaneous coupling technique with Fast Garnet GBC by Pugh and Walker (1961a). Hayashi (1965) synthesized the Naphthol AS-BI substrate and used it with hexazotized pararosanilin in a coupling method. The latter method is the one that is most commonly used at the present time.

Anderson and Leabeck (1961) synthesized 5-bromoindoxyl N-acetyl-β-glucosaminide. This substrate was used either in a dye-coupling method with Fast Garnet GBC or the liberated indoxyl was oxidized to indigo by the addition of potassium ferricyanide to the incubation mixture (Pugh and Walker, 1965; unpublished results). Since by using either the Naphthol AS or the indoxyl technique the enzymatic activity has been localized satisfactorily at the light microscope level, attempts have been made to modify these methods for the study of N-acetyl-β-glucosaminidase at the subcellular level.

Preliminary investigations were carried out with Naphthyl AS-LC N-acetyl-β-glucosaminide. Although the indoxyl technique yields excellent localization, indoxyl substrates are available for only a few enzymes. Naphthol AS substrates, on the other hand, are available for a large number of enzymes. It is an advantage to have a series of techniques in which the same reaction product is formed, because this allows the cytological distribution of enzymes to be compared directly.

Several electron-cytochemical techniques have been developed for Naphthol AS compounds, and these have been fully reviewed by Bowen in this volume. The Naphthyl AS LC substrate was used in a postcoupling technique with p-acetoxymercuric aniline diazotate (Smith and Fishman, 1969) and with hexazotized pararosanilin (Lehrer and Ornstein, 1959) in a simultaneous-coupling method (unpublished results). However, these experiments yielded poor results, and so the possibility of utilizing the 5-bromoindoxyl substrate was examined.

The indoxyl liberated by hydrolysis of 5-bromoindoxyl N-acetyl-β-glucosaminide couples even less readily with hexazotized pararosanilin than does Naphthol AS LC, and so an alternative to a dye-coupling method was sought. It is well known that there are disadvantages in using metal precipitation techniques for electron microscopy (Holt and Hicks, 1962). Although the metals most frequently used impart sufficient contrast for high resolution studies, the precipitates are often crystalline, and fixation and incubation conditions must be carefully selected to prevent diffusion and nonspecific staining. Metal precipitation techniques must be used with caution, although useful information has been obtained on the distribution of phosphatases and sulfatases by using lead methods, and on esterases by employing techniques which yield a gold or copper precipitate.

When sections are incubated with the indoxyl substrate and potassium ferricyanide to demonstrate N-acetyl-β-glucosaminidase, iron is present in the deposit formed at the sites of enzyme activity. However, this precipitate cannot be examined with the electron microscope without further treatment, because it is unstable under the electron beam. This loss of reaction product can be prevented by coupling the ferrocyanide formed during oxidation of indoxyl with cupric ions to form insoluble copper ferrocyanide. This metal precipitate is almost unaffected by the electron beam. A precipitate of copper ferrocyanide has been used for the demonstration of other enzyme activities (Karnovsky and Roots, 1964; Ogawa *et al.*, 1968). It has the advantage of being finely crystalline and does not obscure the underlying cellular structure.

PROCEDURE

1. Fix small blocks of tissue (less than 5 mm thick) in freshly prepared formalin, calcium chloride (Holt, 1958) or in 10% formalin adjusted to pH 6.8 with barbiturate buffer (sodium veronal) for 20 hr at 4°C. Approximately 70% of the enzyme activity survives formalin fixation (glutaraldehyde was not used for fixation because it destroys the greater part of the enzyme activity). Bone marrow or other finely particulate material should be pipetted into a centrifuge tube containing cold fixative and fixed for 15 min, followed by centrifugation at a low speed. The precipitated material should be processed in the same manner as other tissues.

2. Prepare small blocks (0.5 mm³) or slices (0.5 mm thick) from the formalin fixed material. Alternatively, 50 μ thick sections may be cut on the freezing microtome, although the morphology of these frozen sections is inferior to that of the small blocks.

3. Rinse the blocks or sections in three changes of cold (4°C) 0.22 M sucrose. Transfer the blocks to the preheated incubation mixture.

4. Incubate the blocks at 37°C for 30 to 45 min in a mixture consisting of 7.0 ml of an aqueous saturated substrate solution (~ 30 mg 5-bromoindoxyl N-acetyl-β-glucosaminide per 100 ml of water), 1.0 ml of 0.5 M citrate buffer, 1.0 ml of 0.005 M potassium ferricyanide, 1.0 ml of 0.025 M copper sulfate, and 0.75 gm sucrose. The solution should be freshly prepared and filtered prior to use.

The pH of citrate buffers changes on dilution. The final pH of the incubation mixture should be 5.5, which can be obtained by using a 0.5 M citrate buffer of ~ pH 5.1. However, the exact pH required to obtain the desired final pH should be determined by experimentation. Acetate buffers cannot be used, because acetate is a competitive inhibitor of N-acetyl-β-glucosaminidase (Pugh and Walker, 1961b). The potassium ferricyanide solution must be freshly prepared; any ferrocyanide in the solution will form a precipitate with the copper sulfate.

Control incubations in which substrate is omitted or sections are inactivated by heat, should be carried out whenever a fresh tissue is investigated. Negative controls have been obtained on all the tissues examined to date.

5. After incubation, rinse the blocks in cold 0.22 M sucrose and then fix in cold buffered formalin for 5 min. Wash the blocks again in cold sucrose.

6. Transfer the blocks to 1% osmium tetroxide buffered to pH 6.8 with barbiturate buffer at 4°C. Leave the tissue in the osmium tetroxide for 1 to 2 hr, allowing the solution to attain room temperature.

7. After immersion in the buffered osmium tetroxide, dehydrate the blocks rapidly in a series of alcohols, transfer to xylene and embed in Araldite.

8. For light microscopy, soak the formalin fixed blocks in cold gum sucrose freezing mixture (Holt, 1958) for at least 24 hr and section on the freezing microtome. Incubate loose sections in the incubation mixture described above, rinse in water, counterstain with methyl green, and mount in an aqueous medium. The copper sulfate solution may be omitted from the incubation mixture for light microscopy.

Examined with the light microscope, sites of enzyme activity are colored brown by the copper ferrocyanide deposit (Hackett's brown) or blue by indigo if the copper sulfate is omitted from the assay mixture.

NOTES ON PROCEDURE

The formation of the copper ferrocyanide deposit is dependent upon the concentrations of potassium ferricyanide and copper sulfate and the pH of the incubation mixture. When either the dye coupling or the oxidation method is used, a well-localized deposit can be obtained at pH's ranging from 3.5 to 6.0. Sections are usually incubated at pH 4.3, which is the optimum pH for enzyme activity. However, the formation of copper ferrocyanide is pH-dependent (Karnovsky and Roots, 1964), and it has been found that optimum results are obtained at pH 5.5; less reaction product is formed at lower pH's, although the substrate is more rapidly hydrolyzed.

The copper in the incubation mixture couples with the ferrocyanide as it is formed (during the oxidation of indoxyl to indigo by potassium ferricyanide), and this reaction will only occur as the ferrocyanide is first produced. An unsuccessful attempt was made to stain sections with copper sulfate after incubating them in the standard indoxyl ferricyanide mixture; the blue dye precipitate remained unchanged.

The optimum final concentration of potassium ferricyanide in the incubation mixture is 0.005 M. This was investigated by using final concentrations of potassium ferricyanide ranging from 0.002 to 0.008 M. It was found that when a concentration below 0.005 M was used, a diffuse precipitate was obtained; at higher concentrations, the salt caused enzyme inhibition.

The concentration of copper sulfate in the incubation mixture is also critical. The quantity of copper sulfate in the incubation mixture was varied from 0.001 to 0.005 M at pH 5.5. When the concentration of copper sulfate was below 0.002 M, the reaction product was diffuse. In very active tissues (e.g., epididymis) examined with the light microscope, a mixed precipitate of blue indigo and brown copper ferrocyanide was formed occasionally at low copper concentrations. At 0.003 M and above, there was a gradually increased inhibition of enzyme activity; at a concentration of 0.005 M copper sulfate, only a small quantity of precipitate was formed.

The oxidation of indoxyl to indigo, and hence the formation of ferrocyanide ions, occurs more slowly at room temperature than at 37°C. It would seem that this slower reaction should favor the formation of the copper compound. In practice, less precipitate was formed at the lower temperature, and consequently incubation at 37°C is preferred.

LOCALIZATION OF N-ACETYL-β-GLUCOSAMINIDASE

The cytochemical technique described yields a finely crystalline electron-dense deposit at sites of enzyme activity. Control experiments yield negative results. With the possible exception of a secretory tissue (epididymis), little reaction product is lost during dehydration and embedding. The reaction product is localized in discrete particles, which are considered to be lysosomal in every tissue that has been investigated.

The distribution of N-acetyl-β-glucosaminidase activity in rat tissues suggested that the enzyme activity originated in mitochondria, and that lysosomes are formed from these altered mitochondria, often with the addition of endogenous and exogenous materials (Pugh, 1972a and b). The reaction product is localized in lysosomes, many of which contain mitochondrial profiles. Small quantities of reaction product are also found in many mitochondria; the quantity of reaction product and its distribution vary greatly. These results are in agreement with those of Wimborn and Bockman (1968), who suggested that lysosome formation begins with degenerative changes in mitochondria rather than with the concept of a "pure lysosome" produced in the Golgi complex or at another cell site, fusing with a vesicle to form a "secondary lysosome" (De Duve, 1963; De Duve and Wattiaux, 1966).

In all the tissues examined, enzyme activity originated in mitochondria; however, the fate of these mitochondria varied. In the proximal tubules of kidney, the reaction product is localized in droplets ranging from less than 1 μ to over 3 μ in diameter. The large droplet shown in Fig. 4–1a is composed of several smaller enzymatically active bodies, each of which is considered to be

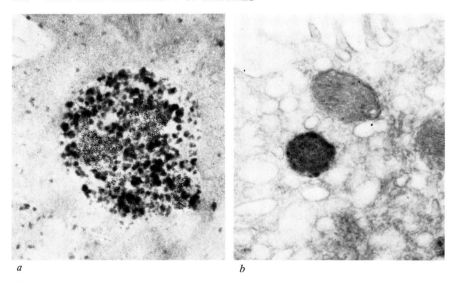

a   b

Fig. 4–1. Rat kidney. *a*: a large lysosome of a proximal tubule cell formed by the aggregation of several smaller lysosomes; *b*: a small lysosome of the distal convoluted tubule. × 40,000.

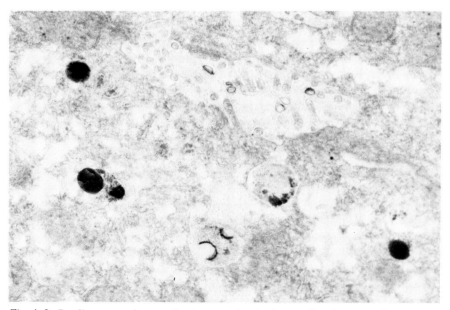

Fig. 4–2. Rat liver parenchyma cells. A group of stained organelles close to a bile caniculus. Two of these peribilary bodies are deeply and homogenously stained; the others were developing full activity. The largest organelle shows a peripheral deposit only; the second body has indentations outlined by the metal precipitate; the third body has two large areas of reaction product in a less active matrix. × 30,000.

derived from an altered mitochondrion. The lysosomes of the distal convoluted segment are smaller (Fig. 4—1b), and form from a single mitochondrion.

The peribilary bodies of rat liver showing N-acetyl-β-glucosaminidase activity (Fig. 4—2) also show evidence of originating from mitochondria; the mitochondria lose their distinctive appearance, becoming shrunken and packed with enzyme activity, probably by elimination of their matrix. This also occurs in gut and connective tissue cells. In epididymis, enzyme activity is found in vesicles and in mitochondria; other cell membranes show slight activity. The mitochondria of this tissue appear to break down and become confluent with the vesicles rather than form separate lysosomes. When viewed with the light microscope, the vesicles are deeply and homogenously stained (by azo dye methods also), whereas with the electron microscope, the center of the vesicles often appears empty. Although enzyme activity may be restricted to the periphery of the vesicles which are intensely active, it is possible that activity has been lost from the vesicles because of a lack of binding sites. Such a possibility of loss of reaction products shows the advisability of combined light- and electron-microscope observations supported, if possible, by biochemical enzyme assays.

The technique presented above can be used for studying blood and bone

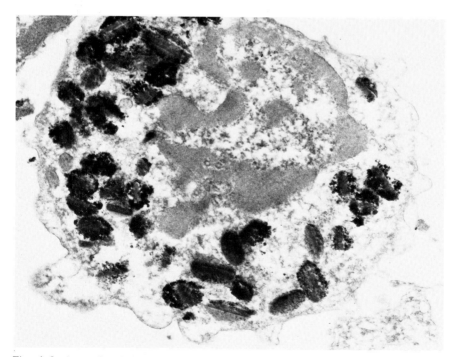

Fig. 4—3. An eosinophil granulocyte from rat bone marrow. The reaction product is localized in the specific granules. × 20,000.

marrow cells and homogenates, provided that a short fixation period is used and the material is handled carefully. An eosinophil leucocyte from rat bone marrow is shown in Fig. 4—3; the reaction product is localized in the prominent specific granules, where it surrounds the discoid crystal.

The distribution of N-acetyl-β-glucosaminidase in plants has not been studied as extensively as it has been in animals. Histochemical methods have been applied to a small selection of plant tissues (unpublished results), and enzyme activity has been found in discrete membrane-bound particles. Acid hydrolases are widely distributed in the plant kingdom, and are associated with a variety of plant structures; spherosomes, aleurone granules, and many vacuoles have lysosomal enzyme activity. Matile (1969) defined plant lysosomes simply as membrane-bound cell components, and advised against making a close comparison between various plant components possessing acid phosphatase activity and animal cell lysosomes. Poux (1963; 1965) localized acid phosphatase in aleurone granules and in small vacuoles of the meristematic cells of *Triticum* shoots and leaves; Hall (1969) demonstrated the enzyme in the spherosomes of the young root tips of several species.

N-acetyl-β-glucosaminidase has a vacuolar distribution. The enzyme is localized in granules arranged in groups in the cells of *Lepidium* roots (Fig. 4—4); some of the granules coalesce to form large stained vacuoles. Enzyme activity is also found in the vacuoles of yeast cells. Yeast cells possess either a single large vacuole or several smaller ones. The reaction product is localized in vacuoles of all sizes and in some very small particles that may or may not be tiny vacuoles (Fig. 4—5). Matile *et al.* (1969) stated that the vacuoles are the lysosomes of yeast cells, and listed several other hydrolytic enzyme activities associated with them.

CONCLUDING REMARKS

The method described here for the localization of N-acetyl-β-glucosaminidase is subject to the usual limitations of electron-cytochemical methods for enzymes. The majority of the enzymes are inhibited to a varied extent by fixation, and the degree of inactivation of N-acetyl-β-glucosaminidase caused by formalin fixation is less than that produced in many hydrolytic enzymes by chemical fixation. Further investigation will be required to determine whether the soluble enzyme activity demonstrated in homogenates (Price and Dance, 1967) is present in living cells, and so inhibited by fixation.

The control experiments show that the method is specific for N-acetyl-β-glucosaminidase. The localization is identical with that of the original dye method and Naphthol AS techniques. The high substantivity and finely crystalline character of the copper ferrocyanide precipitate ensure that the reaction product is suitable for electron microscopy.

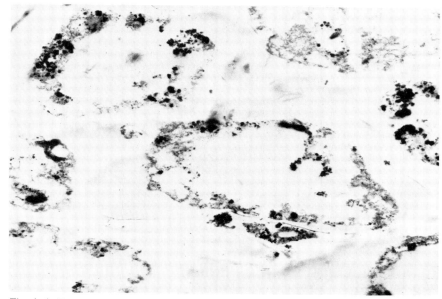

Fig. 4–4. Transverse section immediately behind the piliferous layer of *Lepidium* root. The stain is localized in granules arranged in groups in the cytoplasm; some of these coalesce to form very large stained vacuoles. × 20,000.

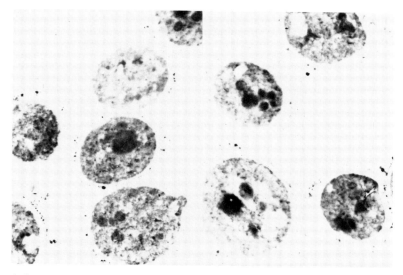

Fig. 4–5. Cells of *Saccharomyces cerevisiae.* The cells contain membrane bound structures of various sizes, which stain deeply with copper ferrocyanide. × 20,000.

It is emphasized that a method which is suitable for localizing an enzyme in one tissue type at high magnifications may not necessarily be suitable for other tissue types. The majority of the cytochemical methods have been devised to yield optimum results with mammalian tissues, and the present technique is no exception. Satisfactory results have been obtained with plant and other animal tissues; using this method, however, it is possible that improvements can be made by modifying the technique for investigations of nonmammalian species. Within these limits, the method can be used to follow changes in enzyme activity or distribution during physiological processes, and to undertake comparative studies.

REFERENCES

Anderson, F. B., and Leabeck, D. H. (1961). Substrates for the histochemical localization of some glycosidases. *Tetrahedron* **12**, 236.

Conchie, J., Findlay, J., and Levy, G. A. (1959). Mammalian glycosidases: distribution in the body. *Biochem. J.* **71**, 318.

De Duve, C. (1963). The Lysosome Concept. In *Lysosomes*. (Reuck, A. V. S., and Cameron, M. P., eds.). J. and A. Churchill, London.

De Duve, C., and Wattiaux, R. (1966). Functions of lysosomes. *Ann. Rev. Physiol.* **28**, 435.

Hall, J. L. (1969). Histochemical localization of β-glycerophosphatase activity in young root tips. *Ann. Bot.* **33**, 399.

Hayashi, M. (1965). Histochemical demonstration of N-acetyl-β-glucosaminidase employing Naphthol AS-BI N-acetyl-β-glucosaminide as substrate. *J. Histochem. Cytochem.* **13**, 355.

Hayashi, M. (1967). Comparative histochemical localization of lysosomal enzymes in rat tissues. *J. Histochem. Cytochem.* **15**, 83.

Helfrich, B., and Iloff, A. (1933). Über Emulsin XIII. Darstellung und fermentative Spaltung von Glykosiden des N-acetyl glucosamins und der 2-Deoxyglucose. *Z. phys. Chem.* **221**, 252.

Holt, S. J. (1958). Indigogenic Staining Methods for Esterases. In *General Cytochemical Methods,* Vol. 1 (Danielli, J. F., ed.), p. 375-398.

Holt, S. J., and Hicks, R. M. (1962). Combination of Cytochemical Staining Methods for Enzyme Localization with Electron Microscopy. In *The Interpretation of Ultrastructure* (Harris, R. J. C., ed.), p. 193-211. Academic Press, London.

Howe, C., and Kabat, E. A. (1953). The action of an enzyme from snail liver and blood group A and O (H) substances. *J. Amer. Chem. Soc.* **75**, 5342.

Humphery, J. H. (1946). Studies on diffusing factors. (2). The action of hyaluronidase preparations from various sources upon some substrates other than hyaluronic acid. *Biochem. J.* **40**, 442.

Karnovsky, M. J., and Roots, L. (1964). A "direct coloring" thiocholine method for cholinesterases. *J. Histochem. Cytochem.* **12**, 219.

Leabeck, D. H., and Walker, P. G. (1957). The preparation and properties of acetochloroglucosamine and its use in the synthesis of 2-acetamido-2-deoxy-β-d-glucosides. *J. Chem. Soc.* **958**, 4754.

Lehrer, G. M., and Ornstein, L. (1959). A diazo coupling method for the electron microscope localization of cholinesterase. *J. Biophys. Biochem. Cytol.* **3**, 399.

Linker, A., Meyer, K., and Weissman, B. (1955). Enzymatic formation of monosaccharides from hyaluronate. *J. Biol. Chem.* **213**, 237.

Matile, P. H. (1969). Plant Lysosomes. In *Lysosomes in Biology and Pathology,* Vol. 1 (Dingle, J. T., and Fell, H. B., eds.), pp. 406-30. North-Holland Publishing Company, Amsterdam.

Matile, P. H., Moor, H., and Robinow, C. F. (1969). Yeast Cytology. In *The Yeasts,* Vol. 1 (Rose, A. H., and Harrison, J. S., eds.), pp. 219-302. Academic Press, London.

Neuberger, A., and Pitt Rivers, R. (1939). The hydrolysis of chitin by an enzyme in *Helix pomatia. Biochem. J.* **33**, 1580.

Ogawa, K., Saito, T., and Mayahara, H. (1968). The site of ferricyanide reduction by reductases within mitochondria as studied by electron microscopy. *J. Histochem. Cytochem.* **16**, 49.

Poux, N. (1963). Cytologie végétale: Sur la presence d'enclaves cytoplasiques en voie de dégénérescence dans les vacuoles des cellules végétales. *C. R. Acad. Sci.* **257**, 736.

Poux, N. (1965). Localisation de l'activite phosphatasique acide et des phosphates dans les grains d'aleurone. *J. Microscopie* **4**, 771.

Price, R. G., and Dance, N. (1967). The cellular distribution of some rat-kidney glycosidases. *Biochem. J.* **105**, 877.

Pugh, D. (1963). The cytology of the digestive and salivary glands of the limpet *Patella. Quart. J. Micr. Soc.* **104**, 23.

Pugh, D. (1964). The cytochemistry of lysosomes, with particular reference to rat kidney. Ph.D. Thesis, London University.

Pugh, D. (1967). The enzyme droplets of vertebrate kidney. *Acta Zool. Stockh.* **48**, 1.

Pugh, D. (1972a). The fine localization of N-acetyl-β-glucosaminidase in rat tissue using an indoxyl substrate. *Ann. Histochem.* **17**, 55.

Pugh, D. (1972b). The formation of large lysosomes in rat kidney. *Cytobios* **5**, 87.

Pugh, D., and Walker, P. G. (1961a). Histochemical localization of β-glucuronidase and N-acetyl-β-glucosaminidase. *J. Histochem. Cytochem.* **9**, 105.

Pugh, D., and Walker, P. G. (1961b). The localization of N-acetyl-β-glucosaminidase in tissues. *J. Histochem. Cytochem.* **9**, 242.

Pugh, D., and Walker, P. G. (1965). The staining of lysosomes in rat kidney. *J. Roy. Micros. Soc.* **84**, 401.

Robinson, D., and Stirling, J. L. (1968). N-acetyl-β-glucosaminidase in human spleen. *Biochem. J.* **107**, 321.

Sellinger, O. Z., Beaufay, N., Jacques, P., Doyen, A., and De Duve, C. (1960). Intracellular distribution and properties of β-N-acetylglucosaminidase and β-galactosidase in rat liver. *Biochem. J.* **74**, 450.

Smith, R. E., and Fishman, W. H. (1969). p(Acetoxymercuric) aniline diazotate: a reagent for visualizing the naphthol AS-BI product of acid hydrolase action at the level of the light and electron microscope. *J. Histochem. Cytochem.* **17**, 1.

Wachtler, K., and Pearse, A. G. E. (1966). The histochemical demonstration of five lysosomal enzymes in the pars distalis of the amphibian pituitary. *Z. Zellforsch.* **69**, 326.

Watanabe, K. (1936). Biochemical studies of carbohydrates. XXII. An animal β-N-monoacetylglucosaminidase. *J. Biochem. Tokyo* **24**, 297.

Wimborn, W. B., and Bockman, D. E. (1968). Origin of lysosomes in parietal cells. *Lab. Invest.* **19**, 256.

5

Glutamate Oxalacetate Transaminase

SIN HANG LEE

Department of Pathology,
Yale University Medical School,
New Haven, Connecticut

INTRODUCTION

Transaminases or amino transferases are enzymes which catalyze the transfer of an amino group from one molecule to another without the intermediate formation of ammonia. The usual reactants involved are ℓ-amino acids and α-keto acids. The overall transamination reaction may be shown as follows:

$$
\begin{array}{ccccccc}
R_1 & & R_2 & & R_1 & & R_2 \\
| & & | & & | & & | \\
CHNH_2 & + & CO & \rightleftharpoons & CO & + & CHNH_2 \\
| & & | & & | & & | \\
COOH & & COOH & & COOH & & COOH
\end{array}
$$

Enzymatic transamination was discovered by Braunstein and Kritzmann (1937), who used preparations of pigeon breast muscle to catalyze the conversion of α-ketoglutarate to ℓ-glutamate in the presence of a number of amino acids serving as amino group donors. Subsequent studies have shown that the scope of amino group transfer reactions is very broad, and not confined to that of ℓ-amino acids and α-keto acids. Transaminases have been found not only in animal tissues but also in plants and microorganisms (Meister, 1962; Guirard and Snell, 1964; Sallach and Fahien, 1969).

The most widely distributed and most extensively studied transaminase is glutamate oxalacetate transaminase (GOT), which catalyzes the following reversible reaction:

$$
\begin{array}{ccccccc}
\text{COO}^- & & \text{COO}^- & & \text{COO}^- & & \text{COO}^- \\
| & & | & & | & & | \\
\text{CH}_2 & & \text{CH}_2 & \text{GOT} & \text{CH}_2 & & \text{CH}_2 \\
| & + & | & \rightleftharpoons & | & + & | \\
\text{CHNH}_2 & & \text{CH}_2 & & \text{CO} & & \text{CH}_2 \\
| & & | & & | & & | \\
\text{COO}^- & & \text{CO} & & \text{COO}^- & & \text{CHNH}_2 \\
& & | & & & & | \\
& & \text{COO}^- & & & & \text{COO}^-
\end{array}
$$

ℓ-aspartate α-ketoglutarate   oxalacetate ℓ-glutamate

This enzyme also bears the names of aspartate aminotransferase and glutamate-aspartate transaminase. The purified GOT has a molecular weight of approximately 110,000, and contains 2 moles of coenzyme, pyridoxal phosphate per mole of enzyme (Jenkins et al., 1959). The pyridoxal phosphate appears to be firmly bound to the apoenzyme. The amino group removed from the amino acids is temporarily retained in the coenzyme, now in the form of pyridoxamine, and will be passed on to its final acceptor, one of the keto acids. Both pyridoxal phosphate and pyridoxamine phosphate have been used as the coenzymes for transaminases, functioning as amino group acceptor or amino group donor, respectively. The formation of these intermediate reaction products is usually disregarded, since in most cases only the overall transamination reaction between the amino acid and the keto acid in the reaction mixture is concerned.

In spite of the fact that there are a large number of transaminases known to exist in almost all biologic systems, only the localization of GOT has been studied to date by investigators interested in enzyme cytochemistry at the subcellular level.

GENERAL CONSIDERATION OF THE CYTOCHEMICAL REACTION

As mentioned above, the GOT reaction system requires two substrates: one amino acid serving as an amino group donor, and one keto acid serving as an amino group receptor. The reaction products also consist of one amino acid and one keto acid, which differ from the substrates only in length of the carbon chain. Since the reaction is reversible, theoretically either of these two pairs of dicarboxylic acids—that is, ℓ-aspartate and ∝-ketoglutarate or ℓ-glutamate and oxalacetate—can be used as substrates for the enzymatic reaction. Besides the substrates, the cytochemical medium must contain a reagent which can capture one of these four organic acids, rendering it insoluble at the site of its formation. In addition, an ideal capturing reagent should satisfy the following requirements:

1. It must not inactivate the enzyme concerned.

2. It must not precipitate the substrates to be used, or prevent them from reacting with the enzyme.

3. It must not initiate any nonenzymatic reaction which may lead to the formation of the final reaction product or to the decomposition of the latter.

4. It should be electron-opaque, or capable of being made so by a secondary chemical reaction without displacing it from its original site of deposition.

5. It must not give rise to a direct differential binding with the tissues or cells during the incubation.

Interestingly enough, the best capturing reagent for the demonstration of GOT activity appears to be lead ion (Pb^{++}), which has already found wide application in the field of enzyme histochemistry since Gomori (1941) introduced it as the precipitant for phosphate released by acid phosphatase. Its use at a physiologic pH has recently generated much controversial discussion (Moses *et al.*, 1966; Novikoff, 1967; Rosenthal *et al.*, 1966 and 1969). It practically fulfills all the above requirements when used under defined conditions. The principle of the technique depends upon, first, a successful chelation of the lead in the cytochemical medium by the substrates and the buffer and, second, the precipitation of a final enzymatic reaction product, oxalacetate by the lead ion (Lee, 1968).

It is not always technically feasible to maintain a high concentration of lead in solution at the physiologic pH without interfering with the capturing of the enzymatic reaction product. Although amino acids are generally known to form chelates with metal ions, the stability of the chelates varies with the molecular structure of the amino acid. In the present system, which contains 6 mM lead nitrate, a final concentration of 20 mM aspartate (ℓ- or d-isomer), a tridentate ligand (Sargeson, 1964), is required to prevent precipitation of the lead salts in solution at a slightly alkaline pH. When the aspartate is replaced with alanine or glutamate on an equimolar basis, the lead will be precipitated as soon as the pH is raised to near neutral range. The type of buffer used in this system is also of great importance. Among all the buffers tested, only imidazole, which was introduced by Mertz and Owen (1940), can be used, and must not be replaced with Tris (hydroxymethyl)-aminomethane, although both were recommended by Gomori (1946) as buffers for histochemical media which contain calcium and require a physiologic pH.

The maintenance of the high stability of lead in solution seems to depend upon the interplay of two essential ingredients, imidazole and aspartate. The latter serves a dual function, being a substrate and chelating agent at the same time. When d-aspartate is used in place of its ℓ-isomer, as in a control medium, no transamination takes place, but the lead will remain stable in solution.

Little has been written about the chemical properties of the heavy metal salts formed by oxalacetate. However, it is well known that some unusual reactions occur between metal ions and oxalacetic acid. For example, oxalacetic acid undergoes nonenzymatic decarboxylation, which is catalyzed by polyvalent cations, leading to the formation of pyruvic acid and carbon dioxide (Krampitz and Werkman, 1941; Krebs, 1942). In addition, as early as 1942, Krebs discovered that, unlike pyruvic and ∝-ketoglutaric acids, oxalacetic acid forms an insoluble precipitate with mercurous ion at an acid pH (Krebs, 1942). However, attempts to use mercurous ion as a capturing reagent for the oxalacetate generated enzymatically were unsuccessful because this metal ion itself is very unstable and tends to form mercurous oxide precipitate at near neutral pH, which is required for the activity of GOT.

Another unusual characteristic of oxalacetate is that the solubility of its lead salt is not substantially influenced by the presence of 20 mM aspartate at pH 7.3, while the solubility of lead pyruvate and lead ∝-ketoglutarate is increased at least ten times under similar conditions (Lee, 1968).

The possibility of using lead ion to precipitate oxalacetate was recorded in an experiment performed by Speck (1949). He tested the effects of more than a dozen polyvalent metal ions on the nonenzymatic decarboxylation of oxalacetic acid at pH 5, and noticed an insoluble white precipitate, perhaps lead oxalacetate in the mixture when the concentration of Pb^{++} reached 0.01 M or greater. The nature of this precipitate can be readily confirmed by the hydrazone method (Lee, 1968). It is always desirable to identify the final reaction product with a biochemical method, since other chemicals, such as lead carbonate and lead oxides, may also form insoluble deposits in the presence of CO_2 and at an alkaline pH. The stability and solubility of lead oxalacetate are highly pH-dependent. Exposure of the tissues or cells to an acid pH should always be avoided once the reaction product is formed. Since lead ion can form precipitates in the presence of Cl^-, HCO_3^-, $H_2PO_4^-$ and HSO_4^-, these anions should be eliminated from all the solutions to be used.

PREPARATION OF THE INCUBATION MEDIUM

Solutions A and B should be made and stored separately.

A.	ℓ-aspartic acid	266.2 mg
	∝-ketoglutaric acid	29.2 mg
	0.25 M sucrose	10.0 ml
	0.2 M imidazole in 0.25 M sucrose	25.0 ml

Dissolve the organic acids completely. Adjust the pH to 7.5 to 7.6 with 0.4 N NaOH. Add up to 50 ml with 0.25 M sucrose. The amount of α-ketoglutaric acid used may be increased to 58.4 mg if a faster rate of reaction is desired. However, for most tissues a slow reaction is preferred.

B.   12 mM lead nitrate in 0.25 M sucrose without buffer

Shortly before use, an equal volume of solution B is added dropwise to solution A with constant stirring or agitation. The final cytochemical medium consists of 20 mM ℓ-aspartate, 2-4 mM α-ketoglutarate, 6 mM lead nitrate, 50 mM imidazole, and 0.25 M sucrose.

Both solutions A and B can be stored separately in cold (2 to 4°C) for at least a week; but they must be warmed up to room temperature before being mixed, because the metal complexes of the organic acids have lower solubility at cold temperatures. The medium when made correctly should be perfectly clear after the lead has been added. No filtration is necessary. The following two control media should be employed:

ℓ-aspartate control

ℓ-aspartic acid	266.2 mg
0.25 M sucrose	10.0 ml
0.2 M imidazole in 0.25 M sucrose	25.0 ml
Dissolve and adjust pH to 7.5 to 7.6	
Add 0.25 M sucrose up to	50.0 ml

d-aspartate control

d-aspartic acid	266.2 mg
α-ketoglutaric acid	29.2 mg
0.25 M sucrose	10.0 ml
0.2 M imidazole in 0.25 M sucrose	25.0 ml
Dissolve and adjust pH to 7.5 to 7.6	
Add 0.25 M sucrose up to	50.0 ml

Equal volume of 12 mM lead nitrate in sucrose is added to both control media before use, as for the GOT cytochemical medium.

The d-aspartate control medium is actually an α-ketoglutarate control. The inactive d-aspartate serves as a chelating agent to keep the lead in solution.

A simple experiment can be performed to test the efficiency of the GOT cytochemical medium:

1. Dilute one drop of purified GOT (Sigma Chemical Co., St. Louis, Mo.) in 5 ml distilled water. The dilute enzyme solution may be stored in a refrigerator at 2 to 4°C. It should not be frozen; otherwise the enzyme protein will be denatured and become insoluble.

2. Transfer one drop of the dilute GOT to a test tube containing 5 ml of the GOT cytochemical medium.

3. Wait for 10 min, during which time the clear medium should gradually turn turbid in the vicinity of the enzyme solution, indicating formation of lead oxalacetate as a result of enzymatic transamination. The medium without the added enzyme should remain clear.

PREPARATION OF THE TISSUE

A considerable proportion of the intracellular GOT is soluble and diffusible. Therefore, only adequately fixed material in which the enzyme molecules are immobilized *in situ* should be used for cytochemical study.

Perfusion of the tissues with a cold fixative via the vascular channels immediately after the blood circulation is interrupted appears to be the best method to preserve the GOT activity in the cells. The flow of perfusate through the vessels should be maintained for 2 to 3 min, and must be rapid enough so that the temperature of the tissue is chilled instantaneously. Therefore a high positive pressure is usually required during perfusion. A 500 ml bottle and a set of tubing used for intravenous administration of fluids, with its air inlet connected to a source of compressed air, are most suitable for perfusion fixation of any organs of the rat via its aorta. Hayat (1970 and 1972) has given detailed procedures for perfusing a wide variety of organs.

The perfusion fluid is a modified Karnovsky's fixative (Karnovsky, 1965), consisting of 1% glutaraldehyde and 3.7% formaldehyde:

0.25 M sucrose	100 ml
Glutaraldehyde (on a 100% basis)	2 ml
	(4 ml if a 50% solution is used)
Formalin (37% formaldehyde)	20 ml
0.2 M imidazole in 0.25 M sucrose	50 ml
Mix and adjust pH to 7.4 with 1N HNO_3	
Add 0.25 M sucrose up to	200 ml

After perfusion, the tissue is cut into small pieces (less than 1 mm^3) while still immersed in the perfusate. Then they are fixed for an additional 30 min in an ice cold 3.7% formaldehyde prepared as follows:

0.25 M sucrose	100 ml
Formalin (37% formaldehyde)	20 ml
0.2 M imidazole in 0.25 M sucrose	50 ml
Adjust pH to 7.4 with 1N HNO_3	
Add 0.25 M sucrose up to	200 ml

The tissue blocks which have been fixed are washed in the following buffered sucrose for at least 2 hr, with multiple changes of solution:

0.25 M sucrose	100 ml
0.2 M imidazole in 0.25 M sucrose	50 ml
Adjust pH to 7.4 with 1N HNO$_3$	
Add 0.25 M sucrose up to	200 ml

The last washing is carried out at room temperature.

It is essential for preservation of the structural-enzymatic relationship to include 0.25 to 0.3 M sucrose in all fixatives and washing fluids. Comparative experiments have shown that mitochondria become swollen and the GOT isozyme normally associated with the cristate membrane is readily inactivated if sucrose is withdrawn from the fixatives or its concentration is reduced to below 0.1 M. Other small molecules do not seem to perform the function of sucrose, since increasing the concentration of imidazole up to 400 mM in the fixatives does not prevent swelling of the mitochondria and the inactivation of the cristate isozyme if sucrose is omitted.

It has been reported that incorporation of a substrate, α-ketoglutarate, in the fixatives preserves more tissue GOT activity during the fixation procedure (Papadimitriou and Van Duijn, 1970a). The reactions between the enzymes and substrates and between the substrates and aldehyde fixatives certainly merit further exploration.

Obviously, it is not always possible to fix tissues by perfusion techniques, especially when human materials are dealt with. Under such circumstances, one must use an immersion method for the initial fixation. For this purpose, fresh tissue blocks should be trimmed to small size and fixed in the cold fixative containing 1% glutaraldehyde and 3.7% formaldehyde in buffered sucrose before they are transferred to the formaldehyde solution. The total duration of fixation in the glutaraldehyde solution should not exceed 5 min. As can be expected with the immersion method of fixation, only one or two superficial layers of intact cells of the tissue block are suitable for cytochemical study. The enzyme activity of the broken cells at the surface of the block is largely inactivated. The cells further toward the center of the block are not fixed by glutaraldehyde.

Occasionally, it is desirable to perform ultracytochemical study of enzymatic activities in the subcellular fraction isolated by differential centrifugation. In order to preserve and demonstrate GOT activity, both cytochemically and biochemically, the subcellular organelles must be fixed while the cells are still intact (Lee and Torack, 1968b and c). Aldehyde fixatives rapidly inactivate the purified GOT (Lee and Torack, 1968a) and the GOT associated with the unfixed isolated subcellular organelles (Lee and Torack, 1968c).

INCUBATION AND POSTINCUBATION TREATMENT

The fixed tissue blocks, after being washed, are incubated in the GOT cytochemical medium and the two control media at room temperature ($25 \pm 3°C$) for 20 to 30 min with constant agitation. At the end of the incubation, they are rinsed in a postincubation washing solution.

For the tissues which have been incubated in the GOT medium or in the ℓ-aspartate control medium, an ℓ-aspartate washing solution should be used and is prepared as follows:

ℓ-aspartic acid	266.2 mg
0.25 M sucrose	10.0 ml
0.2 M imidazole in 0.25 M sucrose	25.0 ml
Adjust pH to 7.3 to 7.4 with 0.4 N NaOH or HNO_3	
Add 0.25 M sucrose to	100.0 ml

For the tissues incubated in the d-aspartate control medium, there may be a considerable amount of α-ketoglutarate still retained in the cells. Therefore, the washing solution should consist of an equimolar d-aspartate, instead of ℓ-aspartate, to avoid a possible enzymatic reaction during the washing procedure. The aspartates in the washing solutions serve as chelating agents to remove most of the lead deposits which may be loosely bound to the tissues. The postincubation rinse, which lasts for about 30 sec, should never be carried out in distilled water or unbuffered sucrose, because the solubility of lead oxalacetate is gradually increased with acidity below pH 7.

Immediately following the postincubation rinse, the tissue blocks are usually postfixed directly for 1 hr in 1% osmium tetroxide in a Veronal buffer, as introduced by Palade (1952), except that the hydrochloric acid used to adjust the pH is replaced by nitric acid to avoid a possible secondary conversion of the reaction product into lead chloride. The pH of the osmium solution should be at 7.3 to 7.4, and must be measured prior to use. Alternatively, the tissue blocks may be immersed in a 1% ammonium sulfide in 0.25 M sucrose for 10 min before postfixation in osmium tetroxide. In the latter case, the final reaction product is converted to lead sulfide, which is more stable than lead oxalacetate. Such a secondary conversion of reaction product may not always be desirable, although there appears to be no significant difference in the localization of GOT activity after the conversion.

DEHYDRATION AND EMBEDDING

After osmication, if the tissue blocks have not been treated in ammonium sulfide, they are rinsed in an imidazole-HNO_3 buffer and dehydrated in an ethanol series which is diluted with the same buffer instead of distilled water. The imidazole-HNO_3 is prepared as follows:

0.2 M imidazole in distilled water	25 ml
1N HNO_3 as required to reach pH 7.4	
Add distilled water up to	100 ml

The tissue blocks are immersed in propylene oxide and embedded in Epon 812 as usual. A wide variety of embedding procedures are given by Hayat in this volume.

INTERPRETATION OF THE CYTOCHEMICAL FINDINGS

The thin sections of the Epon-embedded tissue incubated for GOT activity are usually examined in an electron microscope without further staining or after brief staining with lead citrate (Venable and Coggeshall, 1965). The procedures

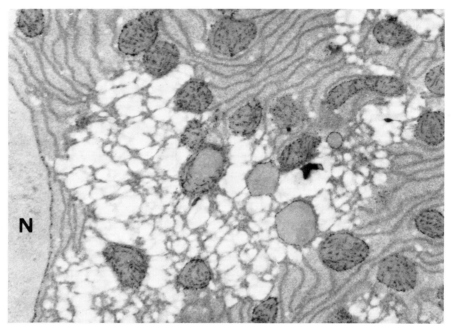

Fig. 5–1. Electron micrograph of rat liver cell. The tissue was initially fixed by perfusion, and incubated in the GOT cytochemical medium for 20 min. The thin section was not counterstained. The electron-opaque enzymatic reaction product was deposited in the mitochondria, at the surface of the microbodies, and in the cisternae of the nuclear envelope. Nucleus (N). × 20,000. From Lee and Torack, 1968b.

for lead staining can be consulted in a recent publication by Hayat (1972). The reaction product appears as electron-opaque deposits, and is localized in the following intracellular organelles of the normal rat tissues:

Liver cell: In the intracristate space and the outer compartment of the mitochondria; at the surface (or in the membrane) of the microbodies; and in the cisternae of the nuclear envelope (Figs. 5—1 and 5—2).

Heart muscle: In the intracristate space and the outer compartment of the mitochondria; in the cisternae of the nuclear envelope; and in the segments of the sarcoplasmic reticulum, including its terminal cisternae (Lee, 1969).

Kidney: In the subapical cytoplasmic vesicles of the distal renal tubules (Lee, 1970); at the surface (or in the membrane) of the microbodies of the proximal tubules (Fig. 5—3); in the intracristate space and the outer compartment of the mitochondria, but not as prominently as in the mitochondria of the liver cell and myocardium; and in the cisternae of the nuclear envelope.

Adrenal cortex: In the tubulocristate-vesicular system of the mitochondria, but not in the outer compartment between the two surface limiting membranes; and in the cisternae of the nuclear envelope (Chak and Lee, 1971).

In addition, GOT activity has also been reported in other murine tissues

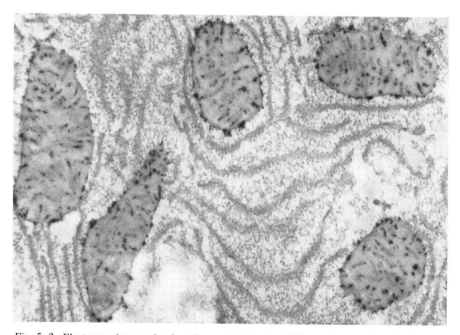

Fig. 5—2. Electron micrograph of rat liver cell, showing GOT reaction product localized in the cristae and probably in the outer compartment of the mitochondria. The thin section was briefly stained with lead citrate. X 37,000. From Lee and Torack, 1968b.

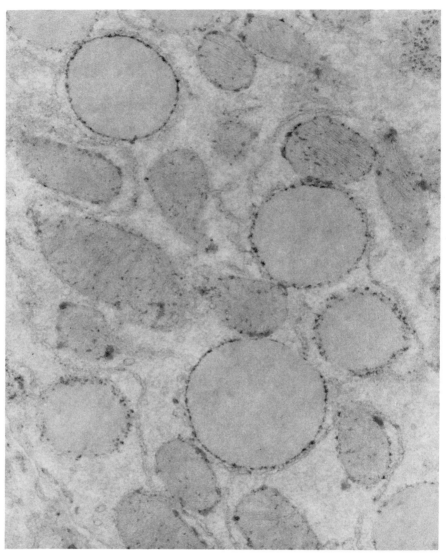

Fig. 5—3. Electron micrograph of proximal renal tubular cell of the rat, showing localization of GOT reaction product at the surface of the microbodies and, less prominently, in the cristae and outer compartment of the mitochondria. The tissue was incubated in the cytochemical medium for 20 min; and the thin section was unstained. × 55,000.

(Papadimitriou and Van Duijn, 1970b) and in the cell membrane of *Toxoplasma gondii* (Akao, 1971).

One must bear in mind that only the reaction product resulting from the activity of an enzyme can be visualized, but not the enzyme molecule itself. The reaction product which accumulates in the outer compartment of the mitochondria may be formed by an enzyme located on the outer surface of the inner limiting membrane, on the inner surface of the outer limiting membrane, or in the space between these two membranes. However, on the basis of the evidence available (Lee and Torack, 1968b; Lee, 1972), the molecules of GOT in the mitochondria of the rat liver cell are probably incorporated in the inner limiting membrane, with their active sites projecting into the outer compartment, since after mechanical damage and a brief period of storage of the cells in sucrose, the reaction product can be demonstrated only in the mitochondrial matrix. This evidence suggests that a secondary transfer of the enzyme molecules from a site adjacent to the outer compartment to the matrix of the mitochondria takes place.

The interpretation of an electron-opaque heavy metal deposit seen on the electron micrographs is not always an easy matter. Students of enzyme cytochemistry who use lead as a trapping reagent to capture the enzymatic reaction product should be aware of at least two possible undesirable mechanisms which can cause deposition of this heavy metal in the tissues but are not related to the enzyme activity concerned.

First, there may be a direct differential binding of lead to the tissues during the incubation. This process is unrelated to any reaction products generated in the cytochemical medium. Such a possibility always exists, but becomes a real misleading phenomenon when unfixed tissues or tissues fixed only briefly in a very dilute (0.3%) glutaraldehyde solution are used, as pointed out by Gillis and Page (1967) and Podolsky (1968). This complication appears negligible with the GOT cytochemical techniques as recommended above, in which an appropriately fixed material is employed and a rather strong chelating agent, aspartate, is used in the medium. However, adequate controls should always be included in all experiments to exclude such a possible nonspecific binding.

Second, in some cases, a nonenzymatic reaction, which may occur spontaneously in the medium or may be initiated by the capturing reagent, can lead to the formation of a product indistinguishable from that of a true enzymatic reaction. In general, biochemists are in a more fortunate position to solve this problem, for the rate of nonenzymatic reaction can always be subtracted from the total simply by adding a "blank" tube with no enzyme added. The cytochemists who have to deal with both the "blank" and the true enzymatic reactions in the same cell or in the same organelle are often bewildered. Needless to say, it becomes difficult at times to decide whether a deposit observed on an electron micrograph is the product of nonenzymatic reaction or that due to a true enzymatic activity, should both exist. Fortunately, with the GOT cytochemical medium, a spontaneous transamination between substrates does not take place, even in the presence of lead. Therefore, the "blank" in this system is practically zero.

Although one can be reasonably certain that all the lead deposits in the cells are in the form of GOT reaction product during the initial period of incubation, which may vary from 10 to 30 min, depending on the level of the enzymatic activity in the cell, complications resulting in nonspecific precipitation of lead do occasionally occur. An early sign of nonspecific precipitate is a haphazard deposition of lead in the nucleoplasm without any relation to the structural organization of the nuclei. This phenomenon has long been recognized by enzyme histochemists who use metals as capturing reagents (Gomori, 1950; Novikoff, 1951; Herman and Deane, 1953). Since it usually occurs only after prolonged incubation or after the incubated tissue blocks have been exposed to solutions with a low pH, it is assumed that a depletion of the chelating agent (aspartate) in the tissue and the increase in solubility of the lead salt of oxalacetate in an acid environment probably play important roles in the formation of these nonspecific deposits.

IDENTIFICATION OF ENZYMATIC REACTION PRODUCT

In order to obtain a better correlation between the cytochemical findings and the biochemical data, sometimes it is necessary to use a biochemical method to identify the end product captured cytochemically in the fixed tissues or in the purified subcellular organelles isolated from them. For this purpose, the techniques are outlined below, and are carried out at room temperature:

The tissue homogenates or the isolated subcellular fractions to be studied are incubated in the GOT cytochemical medium with or without sucrose for 20 to 30 min. The suspension is then centrifuged, and the supernatant discarded. The sediments should be washed twice with 0.05 M imidazole-HNO_3 buffer, and resuspended in 0.1 M phosphate buffer (pH 7.4). In this solution, the oxalacetate is displaced from its lead salt by the phosphate, which forms a lead compound with a much lower solubility than that of lead oxalacetate. After centrifugation, 1 ml supernatant (dilute with 0.1 M phosphate if necessary) is removed and mixed with 1 ml of 1 mM 2,4-dinitrophenyl hydrazine in 1 N HCl. Twenty min later, 10 ml 0.4 N NaOH is added to the mixture. The optical densities of the solution are read in a spectrophotometer at wave lengths from 350 μ to 550 μ continuously and recorded. The absorption spectrum should be typical for the hydrazone of oxalacetic acid (Reitman and Frankel, 1957; Lee, 1968).

REFERENCES

Akao, S. (1971). Toxoplasma gondii: Aspartate aminotransferase in cell membrane. *Exptl. Parasitol.* **29**, 26.

Braunstein, A. A., and Kritzmann, M. G. (1937). Über den Ab-und Aufbau von Aminosäuren durch Umaminierung. *Enzymologia* **2**, 129.

Chak, S. P., and Lee, S. H. (1971). Ultrastructural localization of glutamic oxaloacetic transaminase activity in adrenal cortical cell of the rat. *J. Ultrastruct. Res.* **35**, 265.

Gillis, J. M., and Page, S. G. (1967). Localization of ATPase activity in striated muscle and probable source of artifact. *J. Cell Sci.* **2**, 113.

Gomori, G. (1941). Distribution of acid phosphatase in the tissues under normal and under pathologic conditions. *Arch. Path.* **32**, 189.

Gomori, G. (1946). Buffers in the range of 6.5 to 9.6. *Proc. Soc. Exptl. Biol. Med.* **62**, 33.

Gomori, G. (1950). Sources of error in enzymatic histochemistry. *J. Lab. Clin. Med.* **35**, 802.

Guirard, B. M., and Snell, E. E. (1964). Vitamin B_6 Function in Transamination and Decarboxylation Reactions. In *Comprehensive Biochemistry*. Vol. 15, pp. 138-99. (Florkin, M., and Stotz, E. H., eds.). Elsevier, Amsterdam.

Hayat, M. A. (1970). *Principles and Techniques of Electron Microscopy: Biological Applications*, Vol. 1. Van Nostrand Reinhold Company, New York.

Hayat, M. A. (1972). *Basic Electron Microscopy Techniques*. Van Nostrand Reinhold Company, New York.

Herman, E., and Deane, H. W. (1953). A comparison of the localization of alkaline glycerophosphatase, as demonstrated by the Gomori-Takamatsu method, in frozen and in paraffin sections. *J. Cell. Comp. Physiol.* **41**, 201.

Jenkins, W. T., Yphantis, D. A., and Sizer, I. W. (1959). Glutamic Aspartic Transaminase. I. Assay, purification, and general properties. *J. Biol. Chem.* **234**, 51.

Karnovsky, M. J. (1965). A formaldehyde-glutaraldehyde fixative of high osmolarity for use in electron microscopy. *J. Cell Biol.* **29**, 137A.

Krampitz, L. O., and Werkman, C. H. (1941). The enzymic decarboxylation of oxaloacetate. *Biochem. J.* **35**, 595.

Krebs, H. A. (1942). The effect of inorganic salts on the keto decomposition of oxaloacetic acid. *Biochem. J.* **36**, 303.

Lee, S. H. (1968). Histochemical demonstration of glutamic oxalacetic transaminase. *Amer. J. Clin. Path.* **49**, 568.

Lee, S. H. (1969). Ultrastructural localization of glutamic oxalacetic transaminase activity in cardiac muscle fiber and cardiac mitochondrial fraction of the rat. *Histochemie* **19**, 99.

Lee, S. H. (1970). The possible role of the vesicles in renal ammonia excretion: an implication of concentrated glutamic oxalacetic transaminase. *J. Cell Biol.* **45**, 644.

Lee, S. H. (1972). *Ultracytochemistry of the mitochondrial glutamate oxalacetate transaminase activity*. Proc. 4th Internat. Cong. Histochem. Cytochem. Kyoto, Japan. pp. 107-108.

Lee, S. H., and Torack, R. M. (1968a). Effects of lead and fixatives on activity of glutamic oxalacetic transaminase. *J. Histochem. Cytochem.* **16**, 181.

Lee, S. H., and Torack, R. M. (1968b). Electron microscope studies of glutamic oxalacetic transaminase in rat liver cells. *J. Cell Biol.* **39**, 716.

Lee, S. H., and Torack, R. M. (1968c). A biochemical and histochemical study of glutamic oxalacetic transaminase activity of rat hepatic mitochondria fixed *in situ* and *in vitro*. *J. Cell Biol.* **39**, 725.

Meister, A. (1962). Amino group transfer (survey). In *The Enzymes*. 2d ed. Vol. 6., pp. 193-217. (Boyer, P. D., Lardy, H. A., and Myrbäck, K., eds.). Academic Press, New York.

Mertz, E. T., and Owen, C. A. (1940). Imidazole buffer: its use in blood clotting studies. *Proc. Soc. Exptl. Biol. Med.* **43**, 204.

Moses, H. L., Rosenthal, A. S., Beaver, D. L., and Schuffman, S. S. (1966). Lead ion and phosphatase histochemistry. II. Effect of ATP hydrolysis by lead ion on the histochemical localization of ATPase activity. *J. Histochem. Cytochem.* **14**, 702.

Novikoff, A. B. (1951). The validity of histochemical phosphatase on the intracellular level. *Science* **113**, 320.

Novikoff, A. B., (1967). Enzyme localizations with Wachstein-Meisel procedures: real or artifact. *J. Histochem. Cytochem.* **15**, 353.

Palade, G. E. (1952). A study of fixation for electron microscopy. *J. Exptl. Med.* **95**, 285.

Papadimitriou, J. M., and Van Duijn, P. (1970a). Effects of fixation and substrate protection on the isozymes of aspartate aminotransferase studied in a quantitative cytochemical model system. *J. Cell Biol.* **47**, 71.

Papadimitriou, J. M., and Van Duijn, P. (1970b). The ultrastructural localization of the isozymes of aspartate aminotransferase in murine tissues. *J. Cell Biol.* **47**, 84.

Podolsky, R. J. (1968). Deposit formation in muscle fibers following contraction in the presence of lead. *J. Cell Biol.* **39**, 197.

Reitman, S., and Frankel, S. (1957). Colorimetric method for the determination of serum glutamic oxalacetic and glutamic pyruvic transaminases. *Amer. J. Clin. Path.* **28**, 56.

Rosenthal, A. S., Moses, H. L., Beaver, D. L., and Schuffman, S. S. (1966). Lead ion and phosphatase histochemistry. I. Non-enzymatic hydrolysis of nucleoside phosphates by lead ion. *J. Histochem. Cytochem.* **14**, 698.

Rosenthal, A. S., Moses, H. L., Ganote, C. E., and Tice, L. (1969). The participation of nucleotide in the formation of phosphatase reaction product: a chemical and electron microscope autoradiographic study. *J. Histochem. Cytochem.* **17**, 839.

Sallach, H. J., and Fahien, L. A. (1969). Nitrogen Metabolism of Amino Acids. In *Metabolic Pathways*. 3d ed. Vol. 3, pp. 1-95. (Greenberg, D. M., ed.). Academic Press, New York.

Sargeson, A. M. (1964). Optical phenomena in metal chelates. In *Chelating agents and metal chelates*. (Dwyer, F. P., and Mellor, D. P., eds.), pp. 183-235. Academic Press, New York.

Speck, J. F. (1949). The effect of cations on the decarboxylation of oxalacetic acid. *J. Biol. Chem.* **178**, 315.

Venable, J. H., and Coggeshall, R. (1965). A simplified lead citrate stain for use in electron microscopy. *J. Cell Biol.* **25**, 407.

6

Myrosinase in Cruciferous Plants

TOR-HENNING IVERSEN

Department of Botany,
University of Trondheim,
Trondheim, Norway

INTRODUCTION

Glucosinolates are a structural class of anions in plants of the Rhoedales, which includes the Cruciferae. They are characterized by an enzymatic hydrolysis yielding glucose, sulfate, and isothiocyanates. The enzyme, capable of catalyzing the hydrolysis of glucosinolates (thioglucosides), is known as myrosinase (sinigrinase, β-thioglucosidase, thioglucoside glucohydrolase E.C. 3.2.3.1.). The name "sinigrinase" was given because the best-known substrate is the natural thioglucoside sinigrin (semisystematic name: alyllglucosinolate ion). The general structure of the thioglucosides and the products obtained from the enzymatic hydrolysis are presented below:

$$R-C-S-C_6H_{11}O_5$$
$$\overset{\parallel}{\underset{N-O-SO_3}{}} \xrightarrow{\text{enzyme}} R-N=C=S + C_6H_{12}O_6 + H^+ + SO_4^{2-}$$

(glucosinolate)   (isothiocyanate) (D-glucose)   (sulfate ions)

The aglucone (R) may be any organic group; for sinigrin, R is $CH_2=CH\text{-}CH_2\text{-}$. The enzymatic reaction proceeds in two steps. First, the thioglucosidase liberates aglucone, and then aglucone decomposes to isothiocyanate and sulfate non-enzymatically and spontaneously in neutral solutions (Ettlinger and Kjaer, 1968).

The glucosinolates as a group have been named "mustard oil glucosides," because the common name of isothiocyanate is mustard oil. This name has been found to be inadequate because, upon enzymic hydrolysis, degradation products other than isothiocyanates are formed.

Little is known concerning the function of glucosinolates in plants except that they are precursors of organic isothiocyanates, thiocyanates, and cyanides. The concentration of free isothiocyanates in the intact plant tissue is negligible. They are formed by the cleavage of glucosinolates only after injury of plant tissue, but the role of glucosinolates in catabolism in an intact organism is unknown. A specific type, indole glucosinolate, which occurs in the aerial parts of cruciferous plants, is transformed *in vitro* to indole acetonitrile (Gmelin and Virtanen, 1961). A relationship between glucosinolates and metabolism of growth hormones (e.g., auxins-indole compounds) in plants, however, has not been found (Schraudolf and Bergman, 1965; Schraudolf, 1966).

The occurrence of glucosinolates and myrosinases is closely related in higher plants. In dicotyledonous angiosperms, glucosinolates occur in eleven families (Ettlinger and Kjaer, 1968), the two largest being Cruciferae and Euphorbiaceae. Glucosinolates and myrosinases appear to be present in all species of Cruciferae, but only in two species of Euphorbiaceae.

By using light microscopy, myrosinases have been reported to be present in specialized cells called idioblasts (Guignard, 1890; Schweidler, 1905; Peche, 1913). These cells were confined to the parenchymatous tissue of the green parts of different plants in the family Cruciferae, especially in epidermal cells of leaves (Schweidler, 1905). Detailed studies of ultrastructural localization of myrosinase in the crucifer *Sinapis alba* roots indicate that the enzyme is present in the majority of the root-tip cells and not confined to specific cells (Iversen, 1970a).

In this chapter, the general procedure for localization of myrosinase is presented in detail. Since electron-cytochemical localization of enzymatic activity is limited by problems of methodology, the progress in the development of cytochemical procedures is correlated with biochemical analysis of the enzyme.

The method presented for localization of myrosinase is based upon precipitation of one of the reaction products (sulfate ions) of the hydrolysis of sinigrin. Electron-opaque deposits of lead sulfate are formed at the reaction sites in plant cells when the tissue is incubated with sinigrin in the presence of positive lead ions. Since the enzymatic reaction is influenced by factors such as pH, temperature, fixative, and monovalent cations and anions, the validity of cytochemical techniques is strengthened when the data are correlated with biochemical analysis of the enzyme *in vitro*.

EFFECTS OF ORGANIC AND INORGANIC
SUBSTANCES ON THE ENZYME ACTIVITY

The influence of different substances on crude and purified preparations of myrosinase enzyme isolated from cruciferous seeds has been examined by Tsuruo *et al.* (1967). They prepared myrosinase solutions from yellow mustard seeds by ammonium sulfate precipitation, and studied chromatographic behaviors on ion exchange resins. They found that myrosinase was a single β-thioglucosidase which had a wide pH range, with an optimal peak between pH 6 and 7 when the enzymatic reaction was carried out in citrate-phosphate or Tris buffer.

Studies of the effect of L-ascorbic acid and neutral salts on myrosinase activity showed that the enzyme was activated when the concentration of the acid was higher than 5×10^{-5} M (Tsuruo and Hata, 1967 and 1968a). Salts containing monovalent anions inhibited ascorbic acid-activated enzyme, but did not influence nonactivated enzyme. This effect was found to be due to ionic strength of the salts. Tsuruo and Hata (1968b) also examined the effect of sugars and glucosides on myrosinase activity. An inhibitory effect of glucose on the enzyme was observed at a low concentration of sinigrin. The inhibition was interpreted as competitive. Sucrose, however, did not inhibit the enzymatic activity.

In an attempt to ascertain the sensitivity of the enzyme to different inhibitors or activators, Iversen (1970a) obtained crude enzyme preparations from defatted seeds of white mustard. The enzyme solution was used as such or partially purified by chromatographic separation.

The influence of fixatives on the crude enzyme was examined using a spectrophotometric method developed by Schwimmer (1961). He observed that when myrosinase acts upon sinigrin, which has a maximum absorbancy at 227.5 nm, the absorbancy increases linearly with time. On the basis of these findings, a similar method was employed for estimating hydrolysis of sinigrin in the presence of fixatives. By using a spectrophotometric method, the only reaction product of the sinigrin-myrosinase reaction which absorbs ultraviolet is isothiocyanate. Based on theoretical and experimental considerations, Schwimmer (1961) concluded that the changes in optical density at the absorption maximum of sinigrin demonstrate myrosinase activity.

The influence of 3% glutaraldehyde (unpurified) and 4% formaldehyde in 0.1 M cacodylate buffer (pH 7.2) on myrosinase activity has been studied. On the addition of crude enzyme preparation to sinigrin solution, a rapid decrease in absorbancy at the absorption maximum for potassium salt of sinigrin (222.5 nm) was observed in the first 8 min of the hydrolysis at 21°C (Fig. 6–1, curve 2) (Iversen, 1970a). The hydrolysis started immediately, and was completed after 10 min. The same result was obtained with 4% formaldehyde. When 3% glutaraldehyde was added, the enzymatic reaction was apparently inhibited. As can be

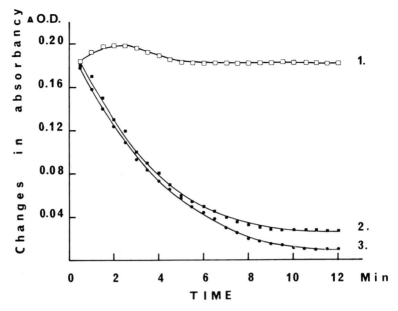

Fig. 6–1. Inhibitory effect of 3% glutaraldehyde on myrosinase activity (curve 1). In the control experiments (curve 2), the absorbancy changes in the reaction mixture start immediately after the addition of myrosinase to the reaction mixture. Addition of 4% formaldehyde slightly stimulates enzymatic activity (curve 3).

seen in Fig. 6–1 (curve 1), a small instantaneous increase was followed by a constant absorbance at 222.5 nm, indicating inhibition of myrosinase. The reasons for the small instantaneous increase are not known.

The inhibition of the enzyme by glutaraldehyde was further tested in experiments with partially purified samples of myrosinase. The crude enzyme was chromatographed on Sephadex G-100 and assayed for thioglucosidase activity estimated by the liberation of sulfate (Fig. 6–2). Inorganic sulfate was determined by a simple method modified after Spencer (1960), or by titration with sodium hydroxide using a recording pH-stat. The peak with the highest myrosinase activity (I in Fig. 6–2), from the chromatographic elution was used to determine the inhibitory effect of 3% purified glutaraldehyde on the enzyme. Under these conditions, the inhibitory effect was 82%, based on the determination of liberated sulfate.

A similar result has been obtained by Wetter (personal communication). Wetter and associates set up a reaction with crystalline sinigrin, purified myrosinase, and cacodylate buffer (pH 7.2), and determined the amount of allyl isothiocyanate released. The experiment indicated that 3% glutaraldehyde treatment resulted in a 84% decrease in activity, while 2% or less did not inhibit myrosinase activity.

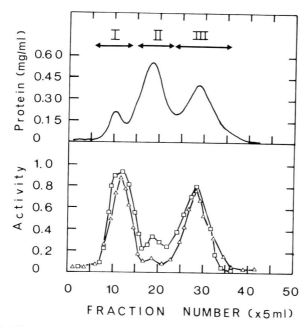

Fig. 6–2. Chromatography of the partially purified myrosinase on Sephadex G-100. The protein applied on the column was eluted with 0.05 M Tris-HCl buffer (pH 7.4) containing 0.05 M NaCl. The fractions (each 5 ml), within the horizontal arrows, were combined and the peaks I and III used for the *in vitro* test to characterize myrosinase. The enzymatic activity is given as liberated sulfate in the presence (□) or absence (△) of 5 × 10⁻³ M L-ascorbic acid. The enzymatic activity is expressed as μmoles/min/ml.

The effect of 4% formaldehyde made up from formalin or generated from paraformaldehyde in cacodylate buffer (pH 7.2) has been tested on purified enzyme solution. All the tests for possible inhibitory effects of formaldehyde showed that in no case was the inhibition greater than 5%, and occasionally, depending upon the substrate concentration, myrosinase activity was slightly stimulated. Observations on the influence of pH, buffers, mono- and divalent anions and cations, and L-ascorbic acid on the enzymatic activity of crude and purified myrosinase samples show that even though the enzyme has a wide pH range, the optimal value lies between pH 6.8 and 7.2. The most desirable buffer appears to be cacodylate.

In test solutions when the ionic strength of the salt was 0.5, cations (K^+, Ca^{++}), anions (Cl^-, SO_4^{--}), and ascorbic acid showed a stimulatory effect on enzymatic hydrolysis. Although ascorbic acid activates the enzyme, it is desirable that this compound and the ions are excluded from the solutions for *in situ* localization of the enzyme. These compounds and ions may have undesirable side reactions with lead nitrate.

Lead nitrate plays an important role in the cytochemical studies on myrosinase (see "General Procedure," below). Therefore the influence of this salt on myrosinase activity has been examined. *In vitro* tests for any possible inhibitory effects of lead nitrate have shown that in no case is the inhibition greater than 10%. Nagashima (1959) observed that crude myrosinase was inhibited 85% by heavy metallic ions such as Hg^{++}, Fe^{++}, Fe^{+++}, and Cu^{++}, but he did not mention the effect of Pb^{++} on the enzyme. Although the aforementioned information on the effects of fixatives and salts on enzymes *in vitro* is significant, it should be noted that enzymes in an isolated state may possess different sensitivity to the reagents from those *in situ*.

EFFECT OF ALDEHYDE FIXATION

It is clear from the above discussion that formaldehyde and glutaraldehyde differ in their ability to preserve myrosinase activity. The usefulness of different aldehydes in electron cytochemistry was first indicated in the detailed studies on rat liver cells by Sabatini *et al.* (1963). They estimated the optimal conditions for fixation, and their results showed that glutaraldehyde inhibits a few enzymes in animal cells. Other studies (Hopwood, 1967; Anderson, 1967) indicated that formaldehyde treatment resulted in better preservation of some enzymes in animal tissues.

Electron cytochemistry of plant tissues has shown that glutaraldehyde is effective in preserving enzymatic activity. After prefixation in glutaraldehyde, different phosphatases remained active in root tips (Poux, 1967; Hall, 1969), in the differentiating secondary vascular cells (Robards and Kidwai, 1969), and in root hairs (Zaar and Schnepf, 1969). Frederick and Newcomb (1969) observed that glutaraldehyde preserved peroxidase activity in mature tobacco leaves. In contrast, very few reports have described the effects of formaldehyde on the activity of plant enzymes at the subcellular level.

Glutaraldehyde is superior to formaldehyde as a fixative for preserving plant ultrastructure. Although glutaraldehyde penetrates into plant tissues more slowly than does formaldehyde, actual fixation is completed more rapidly (Feder and O'Brien, 1968). That speed of fixation is important in the preservation of cellular details has been discussed in depth by Hayat (1970). Although the preservation of cytoplasmic ground substance is less satisfactory after formaldehyde fixation of cruciferous plants, the achievement of specific localization as a result of maintenance of myrosinase activity makes this monoaldehyde preferable for routine preparatory procedures.

GENERAL PROCEDURE

Myrosinase activity can be demonstrated as an insoluble metal salt reaction product of enzymatic hydrolysis. Lead nitrate reacts with sulfate ions resulting from the enzymatic hydrolysis, which results in the production of an insoluble and highly electron-dense-reaction product.

The presence of a high content of myrosinase in white mustard (*Sinapis alba*) makes this species suitable for cytochemical studies. The plants are cultivated from seeds which have been sterilized with 3% hypochlorite for 15 min. The seeds are germinated for 3 days on filter paper in Petri dishes in the dark at 23°C. They are then transferred to a growth chamber with a relative humidity of 70%, and continuously illuminated with fluorescent lamps for 2 days at the same temperature. After this period, myrosinase activity can be demonstrated in different parts of the seedlings.

Prefixation

Tissues from roots, stems, and leaves are cut into small pieces (1 to 2 mm), and central longitudinal slices (less than 0.5 mm in size) are immersed in the fixative. Two types of fixatives are used; 3% glutaraldehyde in 0.1 M cacodylate buffer (pH 7.2), and 4% formaldehyde in 0.1 M cacodylate buffer (pH 7.2). Glutaraldehyde solution is prepared by mixing purified 25% glutaraldehyde with the buffer to a final concentration of 3%.

Stock solutions of glutaraldehyde are affected during storage by changes in pH and temperature, resulting in many impurities. Glutamic acid, which is the final oxidation product of glutaraldehyde and one of the main impurities, can be partially neutralized by adding a small quantity of barium carbonate to the solution. The precipitate which forms is removed by centrifugation, and thus a clear solution is obtained. It is, however, preferable to use a solution of a higher degree of purity and concentration for this type of cytochemical study. The technique described by Fahimi and Drochmans (1965) is recommended. In this technique, glutaraldehyde solution is treated with activated charcoal followed by vacuum distillation. Other methods employed for the purification of glutaraldehyde have been discussed by Hayat (1970).

Stock solutions of formaldehyde, which must be methanol-free, are prepared by dissolving paraformaldehyde powder in twice-distilled water at 60 to 65°C. A few drops of 1 N sodium hydroxide solution are added until the solution becomes alkaline and clear. The final fixative is prepared by diluting the stock solution with cacodylate buffer.

To the fixative is added 1.5% sucrose to balance the tonicity of the solution. The addition of sucrose does not influence myrosinase activity substantially. The selected concentration of sucrose in the fixative has been shown to prevent swelling, and gives a satisfactory preservation of the fine structure. Vacuolated cells in the root or stem present special problems in that they occasionally show disrupted cell membranes and dispersed cytoplasm.

Tissue specimens are prefixed for 2 hr at 4°C and then rinsed for 2 hr with four changes of 0.1 M cacodylate buffer containing 1.5% sucrose.

Incubation

Tissue specimens are incubated in a medium containing glucosinolate (substrate) and lead nitrate. The reaction mixture is prepared immediately before use and contains:

0.1 M sinigrin (potassium myronate, sinigroside)	
in water	0.4 ml
5 mM $Pb(NO_3)_2$ in water	0.3 ml
0.1 M cacodylate buffer (pH 7.2)	0.8 ml
Sucrose (1.5%)	22.5 mg

The pH of the reaction mixture is adjusted to pH 7.2.

Controls must be examined to determine whether the activity can be ascribed to myrosinase and not to artifacts caused by glucosinolate or lead nitrate. The control media contain: the reaction mixture in which distilled water has replaced the substrate, and the reaction mixture in which distilled water has replaced $Pb(NO_3)_2$. The incubation is carried out in small, corked vials for 30 to 60 min at 32°C.

Postfixation, Embedding, and Staining

Following the incubation, tissue specimens are rinsed for 2 hr with four changes of 0.1 M cacodylate buffer (pH 7.2) and postfixed at 4°C for 2 hr in 2% osmium tetroxide made up in the same buffer.

Dehydration is carried out in a graded series of ethanol at 4°C followed by 100% ethanol (two changes, 30 min each) at room temperature. Tissue specimens are transferred to propylene oxide and infiltrated and embedded in Epon. For detailed procedures for embedding, the reader is referred to Hayat (1972).

Ultrathin sections are mounted on copper grids and stained for 15 min with 1% aqueous uranyl acetate. Control sections from the same specimen are double-stained with 1% uranyl acetate for 15 min followed by lead citrate for the same period.

Fig. 6–3. Cytochemical localization of myrosinase activity after prefixation with formaldehyde. *A:* Electron-dense deposits in dilated cisternae (dc) and plasmalemma (p) of outer root cap cells after incubation with 13.2 mM sinigrin for 60 min. w=cell wall. × 40,000. *B:* Positive reaction on the membranes of mitochondria (m), dilated cisternae (arrows), and plasmalemma (p) after incubation with a high concentration of sinigrin (26.3 mM) for 60 min. A slight labeling can also be observed on the nuclear membrane (nm). × 25,000. *C:* Electron deposits on membranes of dilated cisternae (dc, arrows), mitochondria (m), plasmalemma (p) and tonoplast (t) in columella cells of the root incubated with 21.3 mM sinigrin for 30 min. The vacuole (v) and an amyloplast (a) with a starch grain (s) are present. × 30,000. Iversen, 1970a. Used by permission.

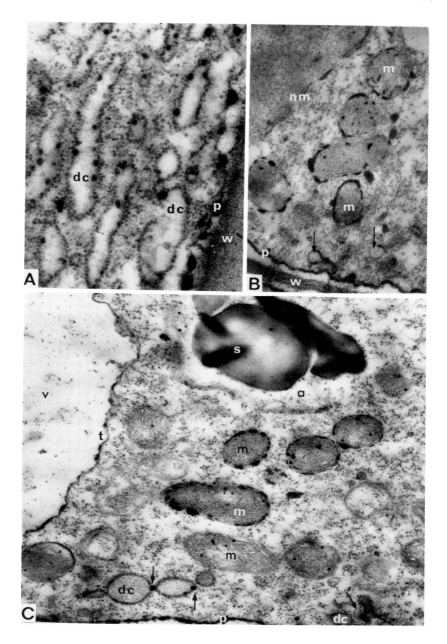

Localization of Reaction Products

Incubation of formaldehyde-fixed root tips in media containing sinigrin and lead nitrate results in the precipitation of electron-dense deposits on membranes of mitochondria, dilated cisternae of endoplasmic reticulum (ER), and nuclear membrane (Fig. 6–3). The precipitate is associated mainly with granular ER cisternal membranes, which suggests that myrosinase is membrane-associated. The precipitate is also found near polyribosomes which can be shown in serial sections to be associated with membranes bounding the nearby cisternae.

Electron-dense deposits of lead sulfate seem to be the sites of myrosinase activity. It is, however, possible that the reaction product (sulfate ions) is diffused and transported along the membranes and/or adjacent ground cytoplasm. It is pointed out that staining of the general ground cytoplasm has never been observed by using this cytochemical method.

In the controls, no lead precipitates were found, either in the organelles or associated with the membranes (Fig. 6–4). At relatively high substrate concentration, a positive reaction occasionally occurred following cacodylate-buffered glutaraldehyde fixation (Fig. 6-4B). After incubation with lower substrate concentrations, glutaraldehyde-prefixed specimens showed no electron-dense deposits irrespective of the length of incubation or temperature. The inhibitory effect of glutaraldehyde on the enzyme seen in the *in vitro* experiments, and the confirmation of this inhibition in the cytochemical studies support the view that these precipitates are enzymatic reaction products.

The electron-dense deposits of lead sulfate appear as a fine precipitate made up of circular or spherical particles. The particles localized at the periphery of mitochondria often form confluent masses. The precipitates clearly contrast against the background, even when the sections have been double-stained with uranyl acetate and lead citrate.

Distribution of Reaction Products

In the seedlings of white mustard, myrosinase activity is confined mainly to the root tissue. Preliminary studies by the author show that myrosinase activity cannot be shown in the epidermis of leaves and stems. On the other hand, light

Fig. 6–4. Inhibition of myrosinase activity (control samples). *A* and *B:* Root tip cells in the columella region incubated with 21.3 mM sinigrin for 60 min after prefixation with glutaraldehyde. *A:* Negative enzymatic reaction in mitochondria (m), endoplasmic reticulum (er), and amyloplast (a) containing starch grains (s). × 30,000. *B:* Electron-dense deposits are observed near the cell wall (w) but not in the amyloplast (a) containing a starch grain (s). × 30,000. *C:* Outer root cap cell prefixed with formaldehyde and incubated with the reaction mixture without lead nitrate. Negative reaction is obvious in mitochondria (m), dictyosome (d), endoplasmic reticulum (er), cell wall (w), and amyloplasts (a) containing starch grains (s). × 30,000. *D:* Prefixation with formaldehyde. Incubation of outer cap cells in the reaction mixture without the substrate. Negative reaction is apparent in dilated cisternae of the endoplasmic reticulum (er), dictyosome (d), and amyloplast (a). × 40,000. Iversen, 1970a. Used by permission.

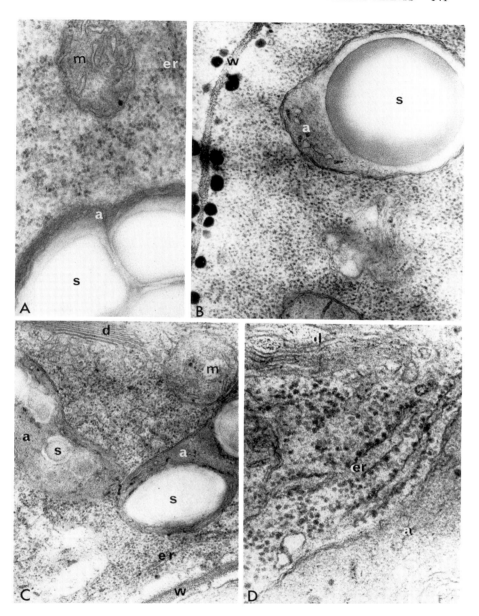

microscopy has shown that myrosinase activity can be localized in idioblasts in parenchymatous tissue of leaves (Schweidler, 1905). Biochemical analysis of myrosinase activity has also suggested the occurrence of the enzyme in the green parts of the plant. Probably methodological limitations have made it difficult to demonstrate cytochemically the enzymatic activity in these parts of the plant. An intensive study of the ultrastructural localization of the enzyme in leaves and stems is desirable.

The enzyme activity is selectively localized in the root cap, the zone of elongation, and the root hair zone. The maximum enzymatic activity is found at the extreme root tip, and seems to decrease proportionately with the distance from the tip. It is not yet clear whether this polar concentration gradient can be ascribed to the experimental procedures or to the natural distribution of the enzyme in the root.

In the regions of root elongation and root hair, the deposits are confined to the cell wall and the cell membrane. Few of the metabolically active organelles (e.g., mitochondria and cisternae of ER) in the vacuolated cells in these parts of the root show positive reaction for the enzyme.

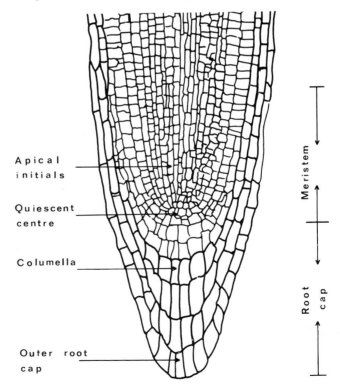

Fig. 6–5. Semischematic map of various zones in the root tip of cruciferous plants.

The root tip of white mustard can be subdivided into four zones (Fig. 6–5). Both normal and dilated cisternae of ER are frequently observed in the outer cap cells, but the number is relatively low in the other regions (Iversen, 1970b). The number of mitochondria is lower in the quiescent center than in other parts of the root tip, which indicates that metabolically active cells are predominant in the latter regions. Myrosinase activity is confined to these two organelles, although, depending upon the experimental procedure, other organelles such as nuclear membrane and amyloplasts may show positive reaction to a limited extent.

The data presented in Table 6–1 show that changes in the incubation period and the concentration of sinigrin can be correlated with the occurrence, distribution, and accumulation of the reaction deposits in different regions and cell organelles. The data presented here are based on experiments in which deposits in several cells from different regions have been scored. The average number of these scores has been expressed as percentage of the total number of organelles present (Iversen, 1970a).

The results presented in Table 6–1 indicate that myrosinase activity is confined to the nuclear membrane, amyloplasts, mitochondria, and dilated cisternae of the ER. Specific localization is dependent upon the composition of the medium, temperature, and duration of incubation. Under optimal conditions, myrosinase activity is confined to dilated cisternae of ER and to mitochondria in meristematic cells.

SOURCES OF ARTIFACTS

The specific localization of myrosinase reaction products in the organelles, the reproducibility of the results, and the control experiments suggest a clear biological basis for the cytochemical method outlined. However, several investigators in this field have emphasized that the localization of enzymatic activity at the subcellular level may be influenced by the fixatives and the composition of the incubation medium. Interpretation of the observations may also be influenced by a possible redistribution of the reaction products and the amount of reaction product deposited.

The difficulties in the correct interpretation of electron-dense deposits in plant and animal tissues have been discussed (Novikoff, 1967; Moses and Rosenthal, 1967; Dauwalder et al., 1969). A discussion of the sources of artifacts in the present method may help to decrease errors in interpretation and perhaps increase the possibility of developing more reliable methodology.

Even if the lead ions inhibit myrosinase activity to some extent (maximum 10%), the concentration of lead ions in the medium is probably sufficiently low not to cause any significant loss of enzyme activity in the tissues of white mustard seedlings. Based on the in vitro experiments, it may also be assumed that the substrate concentration is high enough to saturate myrosinase in vivo.

Table 6–1 Occurrence and Distribution of Myrosinase-Dependent Lead Sulfate Deposits in Root Tips of White Mustard*

Cell type	Outer cap			Columella			Quiescent center			Apical initials		
Substrate concentration (mM)	0.3	13.2	26.3	0.3	13.2	26.3	0.3	13.2	26.3	0.3	13.2	26.3
Cell organelle:												
Nuclear membrane		−	++		−	++		−	+		−	+
Amyloplasts		−	+		+	+		−	++		−	+
Mitochondria		+	++		+	+++		−	+++		−	+
Dilated cisternae	+	+++	+++	+	+++	+++	−	+	++	−	++	++

* The data are selected from Iversen, 1970a. The results presented are based on scores from cells near the median plane of the root tip. The tissue has been prefixed in formaldehyde and incubated with different final concentrations of sinigrin (substrate) for 60 min at 32°C. The number of plus and negative signs indicates:
  − 0 to 5% of the total number of organelles in the cell
  + 5 to 35% of the total number of organelles in the cell
  ++ 35 to 75% of the total number of organelles in the cell
  +++ 75 to 100% of the total number of organelles in the cell

Thus, localization of myrosinase in the root tip and not in other regions of mustard seedling could imply high levels in the root, while the levels in the other plant parts are below the limits of detection by the method. The lower activity in the quiescent center of the root tip (Table 6–1) is in accordance with the low overall rates of metabolic activity in these cells (Jensen, 1957 & 1958; Clowes, 1961).

Artifacts ascribed to overincubation can be caused by too high a substrate concentration. Table 6–1 shows that when the highest substrate concentration is used, there is an extremely heavy precipitation of lead in mitochondria and ER cisternae, while only scattered precipitates are found in both nuclear membrane and amyloplasts (Fig. 6–3). Overincubation can be clearly demonstrated by observing root tips incubated for more than 90 min at a high substrate concentration. A rather generalized lead precipitate is then found in most of the organelles.

Whether the inhibition of myrosinase after prefixation with glutaraldehyde is complete or not is a matter of interpretation. The slight reaction of the cell wall structures after glutaraldehyde fixation (Fig. 6-4b) may represent fixation artifacts.

Although fixation of the ground cytoplasm and organelles is less satisfactory with formaldehyde than that obtained with glutaraldehyde, use of the former fixative is preferable for specific localization of myrosinase activity. The ideal approach, as pointed out by Hayat (1970 and in this volume), would be to develop a mixture of prefixatives which give satisfactory preservation of both the enzymatic activity and the fine structure of the cell. A mixture of low concentrations of paraformaldehyde and glutaraldehyde ($< 2\%$) has not yet been given sufficient attention; such formulations may yield the desired results.

Postfixation with osmium tetroxide and double staining of thin sections may cause loss of reaction products. The method described above includes all these steps, and the possibility that the preparations do not yield a completely accurate representation of myrosinase activity cannot be excluded. The total absence of lead sulfate deposits in the basal parts of the root and the stem may therefore be due to deficiencies of the cytochemical procedure. To facilitate the localization of enzymatic reaction products and to avoid removal of lead sulfate deposits, omission of postosmication and double staining during the processing of a few control sections is suggested.

ALTERNATIVE PROCEDURES

Since cytochemical methods generally have limitations, validity of the results should be tested by other methods. To the author's knowledge, there are no reports of alternative methods for the localization of myrosinase activity at the subcellular level. Ideally, either the enzyme or its reaction products must be localized *in situ*. Confirmation of the results obtained by using cytochemical procedures is, however, possible by using cell fractionation or autoradiography (see Salpeter and Bachmann, 1972).

The usefulness of differential centrifugation in cell biology has long been established. The purity of the separated cell fractions can be determined by their degree of homogeneity when examined with the electron microscope (Matile, 1968; Morré, 1970; Deter, 1973). Matile isolated vacuoles from maize root tips to demonstrate lysosomal enzyme activity in these organelles. Morré incorporated various ^{14}C-labeled substrates into dictyosomes and other cell fractions of onion stems. The fractions were obtained (after incubations for varying periods) from homogenates stabilized with glutaraldehyde. The majority of the organelle fractions showed a high degree of purity, and the extent of contamination of the fraction could be estimated by electron microscopic analysis.

The above approach is also useful for localizing myrosinase activity. This can be carried out in two ways: incubation of the plant tissue in radioactive myrosinase substrates and determination of radioactivity in various cell organelle fractions, and determination of myrosinase activity in isolated cell organelles.

In the first case, it is necessary to have a radioisotopically labeled substrate. Unfortunately, neither [14]C- nor ^{3}H-labeled glucosinolates are available commercially. It has, however, been shown that when seedlings of white mustard are incubated with ^{35}S-sulfate, the label is incorporated into three different glucosinolates (glucobrassicin, neoglucobrassicin, and sinalbin) after a short incubation period (Bergman, 1970).

The above discussion suggests that it may be possible to use radioactively labeled tissue directly or to extract the labeled glucosinolates and then incorporate the compound into other cruciferous plants. After incorporation of the marker and lead nitrate and homogenization of the tissue (using the method suggested by Matile (1968) or Morré (1970) or a combination of the two), it should be possible to localize myrosinase activity in specific organelles. It is assumed in this method that the localization of radioactivity in the organelles represents myrosinase reaction site where ^{35}S has been precipitated as labeled lead sulfate.

Alternatively, tissue specimens could be stabilized with paraformaldehyde, and, after homogenization and organelle separation, myrosinase activity is determined in various organelles. By using this method, the author has shown that the highest myrosinase activity was found in the cell fraction containing normal or dilated cisternae of the endoplasmic reticulum. It is recommended, when using this method, to release membrane-bounded myrosinases by ultrasonic treatment or by exposing to a weak detergent for a few minutes before enzymatic analysis.

Electron autoradiographic analysis of labeled glucosinolates may give valuable information regarding the localization of myrosinase reaction sites. This method has not been used to its fullest extent. One of the reasons for this is that the labeled compound must not be water-soluble or else it will be extracted from the tissue specimen during processing. The ideal isotope for autoradiography is ^{3}H, but, as mentioned above, ^{3}H-labeled glucosinolates are not available commercially. It is assumed, however, that there are no technical problems in preparing, for example, tritiated sinigrin, barring the expense. Other isotopes have also been used, and it is possible that ^{35}S, which has been used by Pelc et al. (1961), may be valuable for autoradiographic analysis of myrosinase.

POTENTIAL RESEARCH AREAS

The physiological significance of glucosinolates and myrosinase in plants is not clear. It is well known that plants which contain large quantities of glucosinolates have a characteristic taste and odor. The fact that isothiocyanates have antibiotic properties (Virtanen, 1962; Wagner et al., 1965) has led to the speculation that glucosinolates make the plant more resistant to attacks by animals or microorganisms. In this connection, studies by Stahman et al. (1943) are of interest. These authors studied the extent of glucosinolates and myrosinase activity in cruciferous plants and their relationship to resistance to clubroot. Although they did not find any correlation, it is desirable to undertake a detailed ultrastructural study of the infected zone in plants inoculated with *Plasmodiophora brassica* (the causal organism of clubroot) and correlate this with cytochemical localization of myrosinase.

Plants containing myrosinase are normally assumed to contain glucosinolates. In a few cases, however, plants not containing glucosinolates have been reported to show myrosinase activity (e.g., in wheat and clover [Kutaček, 1964]). In one family, Flacourtiaceae, the isolation of enzyme preparations that hydrolyze allyl glucosinolates to isothiocyanates has been reported, but no glucosinolates have been found in the same plants (Ettlinger and Kjaer, 1968). Whether these examples are exceptions to the rule, or should be considered as a taxonomical problem, is a matter for further study. Here and in other cytotaxonomical studies, the described method for the localization of myrosinases at the subcellular level may prove to be a valuable tool.

REFERENCES

Anderson, P. J. (1967). Purification and quantitation of glutaraldehyde and its effect on several enzyme activities in skeletal muscle. *J. Histochem. Cytochem.* **15**, 652.

Bergman, F. (1970). Die Glucosinolat-Biosynthese im Verlauf der Ontogenese von *Sinapis alba* L. *Zeitschr. f. Pflanzenphysiol.* **62**, 362.

Clowes, F. A. L. (1961). *Apical Meristems,* Vol. II. Blackwell, Oxford, England.

Dauwalder, M. W., Whaley, W. G., and Kephart, J. E. (1969). Phosphatases and differentiation of the Golgi apparatus. *J. Cell Sci.* **4**, 455.

Deter, R. (1973). Electron Microscopic Evaluation of Subcellular Fractions Obtained by Ultracentrifugation. In *Principles and Techniques of Electron*

Microscopy: Biological Applications, Vol. 3 (Hayat, M.A., ed.). Van Nostrand Reinhold Company, New York.

Ettlinger, M., and Kjaer, A. (1968). Sulfur Compounds In Plants. In: *Recent Advances in Phytochemistry,* Vol. 1 (Mabry, T. J., Alston, R. E., and Runcckles, V. C., eds.). North-Holland Publishing Company, Amsterdam.

Fahimi, H. D., and Drochmans, P. (1965). Essais de standardisation de la fixation de la concentration du glutaraldéhyde. I. Purification et détermination de la concentration du glutaraldéhyde. *J. Microscopie* **4**, 725.

Feder, N., and O'Brien, T. P. (1968). Plant microtechnique: some principles and new methods. *Am. J. Bot.* **55**, 123.

Frederick, S. E., and Newcomb, E. H. (1969). Cytochemical localization of catalase in leaf microbodies (peroxisomes). *J. Cell Biol.* **4**, 343.

Gmelin, R., and Virtanen, A. I. (1961). Glucobrassicin der Precursor von SCN⁻, 3-Indolylacetonitril und Ascorbigen in *Brassica oleracea* species. *Ann. Acad. Sci. Fennicae A.* **II.** *Chemica* No. 107.

Guignard, L. (1890). Sur la localisation des principes actifs dans la graine des Cruciféres. *C.R. hebd. Séances Acad. Sci.* **III**, 920.

Hall, J. L. (1969). Localization of cell surface adenosine triphosphatase activity in maize roots. *Planta* **85**, 105.

Hayat, M. A. (1970). *Principles and Techniques of Electron Microscopy: Biological Applications,* Vol. 1. Van Nostrand Reinhold Company, New York.

Hayat, M. A. (1972). *Basic Electron Microscopy Techniques.* Van Nostrand Reinhold Company, New York.

Hopwood, D. (1967). Some aspects of fixation with glutaraldehyde: a biochemical and histochemical comparison of the effects of formaldehyde and glutaraldehyde fixation on various enzymes and glycogen, with a note on penetration of glutaraldehyde into liver. *J. Anat.* **101**, 83.

Iversen, T. H. (1970a). Cytochemical localization of myrosinase (β-thioglucosidase) in root tips of *Sinapis alba.* *Protoplasma* **71**, 451.

Iversen, T. H. (1970b). The morphology, occurrence, and distribution of dilated cisternae of the endoplasmic reticulum in tissues of plants of the Cruciferae. *Protoplasma* **71**, 467.

Jensen, W. A. (1957). The incorporation of C^{14}-adenine and C^{14}-phenylalanine by developing root tip cells. *Proc. Nat. Acad. Sci.* **43**, 1838.

Jensen, W. A. (1958). The nucleic acid and protein content of root tip cells of *Vicia faba* and *Allium cepa.* *Exp. Cell Res.* **14**, 575.

Kutáček, M. (1964). Glucobrassicin a potential inhibitor of unusual type affecting the germination and growth of plants: mechanism of its action. *Biol. Plant.* **6**, 88.

Matile, P. (1968). Lysosomes of root tip cells in corn seedlings. *Planta* **79**, 181.

Morré, D. J. (1970). *In vivo* incorporation of radioactive metabolites by Golgi apparatus and other cell fractions of onion stem. *Plant Physiol.* **45**, 791.

Moses, H. L., and Rosenthal, A. S. (1967). On the significance of lead-catalysed hydrolysis of nucleoside phosphatases in histochemical systems. *J. Histochem. Cytochem.* **15**, 354.

Nagashima, Z. (1959). Myrosinase. IV. Examinations on the inhibition of myrosinase. *Nippon Nogeikagaku Kaishi* **33**, 881.

Novikoff, A. B. (1967). Enzyme localization with Wachstein-Meisel procedures: Real or artifact. *J. Histochem. Cytochem.* **15**, 353.

Peche, K. (1913). Mikrochemischer Nachweis des Myrosins. *Ber. dtsch. bot. Ges.* **31**, 458.

Pelc, S. R., Combes, J. D., and Budd, G. C. (1961). On the adaptation of autoradiographic techniques for use with the electron microscope. *Exp. Cell Res.* **24**, 192.

Poux, N. (1967). Localisation d'activites enzymatiques dans les cellules du méristème radiculaire de *Cucumis sativus* L. I. Activitès phosphatasiques neutres dans les cellules du protoderme. *J. Microscopie* **6**, 1043.

Robards, A. W., and Kidwai, P. (1969). Cytochemical localization of phosphatase in differentiating secondary vascular cells. *Planta* **87**, 227.

Sabatini, D. D., Bensch, K., and Barrnett, R. J. (1963). Cytochemistry and electron microscopy: The preservation of cellular ultrastructure and enzymatic activity by aldehyde fixation. *J. Cell Biol.* **17**, 535.

Salpeter, M. M., and Bachmann, L. (1972). Autoradiography. In *Principles and Techniques of Electron Microscopy: Biological Applications*, Vol. 2 (Hayat, M. A., ed.). Van Nostrand Reinhold Company, New York.

Schraudolf, H. (1966). Der Stoffwechsel von Indolderivaten in *Sinapis alba* L. I. Synthese und Umsetzung von L-Tryptophan in Etiolierten Hypokotylsegmenten nach Applikation von Indol-2-^{14}C. *Phytochem.* **5**, 83.

Schraudolf, H., and Bergmann, F. (1965). Der Stoffwechsel von Indolderivaten in *Sinapis alba* L. II. Untersuchungen zu Biogenese und Umsetzung von Indolglucosinolaten mit Hilfe von Ringmarkierten C^{14}-Tryptophan und S^{35}-Sulfate. *Planta* **67**, 75.

Schweidler, J. H. (1905). Die systematische Bedeutung der Eiweiss-oder Myrosinzellen der Cruciferen nebst Beiträgen zu ihrer anatomisch-physiologischen Kenntnis. *Ber. dtsch. bot. Ges.* **23**, 274.

Schwimmer, S. (1961). Spectral changes during the action of myrosinase on sinigrin. *Acta Chem. Scand.* **15**, 535.

Spencer, B. (1960). The ultramicro determination of inorganic sulfate. *Biochem. J.* **75**, 435.

Stahman, M. A., Link, K. P., and Walker, J. C. (1943). Mustard oils in Crucifers and their relation to resistance to clubroot. *J. Agri. Res.* **67**, 49.

Tsuruo, I., and Hata, T. (1967). Studies on the myrosinase in mustard seed. Part II. On the activation mode of the myrosinase by L-ascorbic acid. *Agri. Biol. Chem.* **31**, 27.

Tsuruo, I., and Hata, T. (1968a). Studies on the myrosinase in mustard seed. Part III. On the effects of neutral salts. *Agri. Biol. Chem.* **32**, 479.

Tsuruo, I., and Hata, T. (1968b). Studies on the myrosinase in mustard seed. Part IV. Sugars and glucosides as competitive inhibitors. *Agri. Biol. Chem.* **32**, 1420.

Tsuruo, I., Yoshida, M., and Hata, T. (1967). Studies on the myrosinase in mustard seed. Part I. The chromatographic behaviors of the myrosinase and some of its characteristics. *Agri. Biol. Chem.* **31**, 18.

Virtanen, A. I. (1962). Organische Schwefelverbindungen in Gemüse-und Futterpflanzen. *Angew. Chem.* **74**, 374.

Wagner, H., Horhammer, L., and Nufer, H. (1965). Zur Dünnschichtchromatographie von Senfölen und Senfölglukosiden. *Arzneim. Forsch. (Drug Res.)* **15**, 453.

Zaar, K., and Schnepf, E. (1969). Membranfluss und Nucleosiddiphosphatase Reaktion in Wurzelhaaren von *Lepidium sativum*. *Planta* **88**, 224.

7

Enzyme Immunocytochemistry

LUDWIG A. STERNBERGER

Basic Sciences Department, Medical Research Laboratory,
Edgewood Arsenal, Maryland, and Department of Microbiology,
Johns Hopkins University School of Medicine
Baltimore, Maryland

INTRODUCTION

Immunocytochemistry employs the specificity of antibodies for the detection of cell components that bear antigenic determinants. Cytochemical reagents are employed as detectors to visualize the antigen-antibody reaction. Classically, the detector has been incorporated into the antibody molecule proper as a covalent label (Coons, 1961). More recently, it has been found that specificity and sensitivity are enhanced if only unlabeled antibodies are used in a procedure in which the detector is attached to antibody via immunological bonds only (Sternberger, 1969).

Sensitivity is essential to immunocytochemical methods: a solution which is colored to the naked eye may be transparent in a thin layer viewed in the light microscope; deposits of osmium that stain a cell black for light microscopy may be invisible when only a few molecules are viewed in the high-resolution electron microscope. Fluorescent antibody labeling (Coons and Kaplan, 1950) was successful because light emission against a dark (zero emission) background is more readily detected than colored (azo dye-coupled) antibody viewed as the difference of its own light emission from that of a bright background. Ferritin is a good choice for a sensitive marker of antibody for electron microscopy (Singer and Schick, 1961) because its high iron content is concentrated in characteristic electron-opaque micelles that impart an easily discernible structure to the molecule.

Fluorescein (von Mayersbach, 1966), ferritin (Singer, 1959; Pierce *et al.*, 1964), mercury (Kendall, 1965), uranium (Sternberger *et al.*, 1966b), osmium (Sternberger *et al.*, 1966a), and ferrocene (bis-pentadienyl iron) (Franz, 1968) are fixed labels for antibodies that bear stoichiometric relationships to the number of antibody molecules employed. Hence, *total* contrast at low resolution is limited by the number of antibody molecules specifically reacting with antigen, and is therefore dependent upon both the concentration of labeled antibody and of tissue antigen.

The overall contrast of ferritin antibodies is so low that at low resolutions ferritin antibody-labeled tissue components are not detectable, even though at high resolution individual sites can be discerned. On the other hand, metal-labeled antibodies, while discernible and measurable (Sternberger *et al.*, 1970b) at relatively low electron-microscopic resolution, escape detection at the crucial high resolution level necessary for discerning individual antigenic sites or molecules.

The introduction of enzyme markers appears to avoid the above-mentioned difficulties. In one approach, enzyme is covalently linked to the antibody (Nakane and Pierce, 1966; Avrameas and Uriel, 1966). In the other approach, the enzyme is bound to specific antibody solely by immunobiologic processes (Sternberger, 1969). In both approaches, following attachment of the enzyme to the antigen site *via* specific antibody, a cytochemically detectable reaction for the enzyme is carried out with suitable enzyme substrate, yielding thereby a large number of visible product molecules per single molecule of antibody. The present review concerns principles, procedures, and applications of these two approaches for use of enzymes in electron-microscopic immunocytochemistry.

When the sensitivity of a method is increased, special attention must be paid to the avoidance of concomitant decreased specificity. No advantage would be gained, for example, if sensitivity is increased a thousandfold while background reading is also increased a thousandfold. There are two sources of nonspecificity in an immunocytochemical reaction. One is methodologic nonspecificity. It includes background reading and nonspecific binding of the reagents, their by-products and contaminants.

The other source is immunologic nonspecificity due to cross-reaction of specific antibodies with antigenic determinants other than those of intended localization. This source is also due to the presence in antiserum of other species of antibodies specific to antigenic determinants other than those of intended localization. These other species of antibodies may be due to impurities in the antigen preparation used in immunization, to previous antigenic experience of the immunized animals, or to natural antibodies (specific reactivity of immunoglobulin (Ig) not evoked by antigen). An example of cross-reactions of specific antibodies is the localization of human Ig with antiserum to guinea pig Ig (Henle and Henle, 1965), a reagent often used in indirect methods of immunocytochemistry (Coombs *et al.*, 1945; Fife and Muschel, 1959).

Immunodiffusion and immunoelectrophoresis (Crowle, 1961), especially when combined with radioimmunoassay (Yagi *et al.*, 1963), are considered sensitive tests for the minimal number of antigen molecules of different specificities reacting with an antibody preparation. Sensitivities of 0.003 μg antigen nitrogen per ml can be attained (Yagi *et al.*, 1963; Terry and Fahey, 1964; Minden *et al.*, 1967). Since this sensitivity is less than that attainable by the newer enzyme immunocytochemical methods, these tests should be interpreted with caution when used as the sole basis for selection of monospecific antisera in electron immunocytochemistry. Indeed, one should be aware that by enzyme immunocytochemistry both antigens and antibodies may be detected which hitherto have escaped attention.

Examples of methodologic nonspecificity due to background staining are the reaction of microbodies of liver and kidney with diaminobenzidine (DAB) (Novikoff and Goldfischer, 1969; Goldfischer and Essner, 1970) and the presence of endogenous peroxidase in monocytes and macrophages (Van Furth *et al.*, 1970). Nonspecific attachment of reagents includes reaction of free fluorescein, ferritin, or horseradish peroxidase (PO) with the tissue as well as nonspecific attachment of proteins conjugated with these markers, especially when large amounts of marker are used (Goldstein *et al.*, 1961; von Mayersbach, 1967; Vogt and Kopp, 1965a and b; Booyse *et al.*, 1971b).

Since increased sensitivity of enzyme immunocytochemistry is likely to magnify nonspecificity, special attention had to be paid to the development of reagents exhibiting minimal methodologic side reactions and to staining procedures that avoid the majority of sources of immunologic nonspecificity. Specificity is therefore emphasized in the discussion on chemical basis and staining application of the immunocytochemical methods reviewed in this chapter.

ENZYME-CONJUGATED ANTIBODY METHODS

When enzymes are covalently coupled to antibodies, a conjugate can be obtained which has both antibody specificity and enzymatic activity. The criteria for choice of the enzyme include availability in pure form, a cytochemical method for detecting the enzyme reaction product with light and electron microscopy, and absence, or at least sparse distribution, of endogenous enzymes of specificities similar to that of the enzyme chosen for conjugation with the antibody in the examined tissue.

The first enzyme conjugated with immunoglobulin for immunocytochemistry that fulfilled these criteria has been acid phosphatase (Ram *et al.*, 1966). Subsequent introduction of peroxidase (Nakane and Pierce, 1966; Avrameas and Lespinats, 1967) decreased the variability in the results obtained. The reliability and resolution of the electron cytochemical detection method for PO (Graham and Karnovsky, 1966) are surpassed by few other enzyme cytochemical methods. Glucose oxidase has also been used as an enzyme for conjugation with antibodies for light (Avrameas and Lespinats, 1967) and electron microscopy (Kuhlmann and Avrameas, 1971; Seligman, 1971).

For the conjugation of antibodies to enzymes, bifunctional amino acid reagents are employed. Some of these compounds possess two identical reactive groups. For instance, difluorodinitrodiphenyl sulfone (FNPS) (Nakane and Pierce, 1967) contains two fluoronitrophenyl groups, each of which can react with ϵ-amine of lysine or N-terminal amine in protein. Glutaraldehyde, a dialdehyde, reacts with the same groups (Sabatini *et al.,* 1963; Quiocho and Richards, 1964; Richards and Knowles, 1968; Avrameas, 1969). Carbodiimides (Avrameas and Uriel, 1966; Nakane and Pierce, 1966) cross-link proteins by conjugating free carboxyl with free amino groups. With carbodiimides there are few side reactions, and hence the conjugation could conceivably be monitored by measurement of acid consumed. However, this has not yet been accomplished in immunocytochemical applications. For conjugation proper, enzyme and antibodies are reacted with bifunctional reagents in mixtures containing these three components.

Preparation of Peroxidase-Antibody Conjugate with p,p′-Difluoro-m,m′-dinitrodiphenyl Sulfone (FNPS)

In the procedure preferred by Nakane and Pierce (1967), 50 mg of a specific antibody-containing Ig fraction and 50 mg of PO were dissolved in the cold in 0.5 M carbonate buffer at pH 10.0. Under gentle agitation, 0.25 ml of a 0.5% solution of FNPS in acetone were added. Agitation was continued for 6 hr at 4°C. Following dialysis with phosphate-buffered saline (PBS) and centrifugation for removal of a precipitate, the solution was passed through a Biogel P300 column, 41 cm in length, equilibrated with PBS. Three overlapping effluent peaks were obtained from the column when protein concentration was monitored by absorbance at 280 mu and PO by absorbance at 403 μ. The first peak appeared at the void volume of 67 ml, the second peak at 102 ml, and the third peak at 145 ml.

Conjugated material possessing both antibody and enzyme activity as detected by specific staining capacity was found between elution volumes 81 and 105 ml, with a maximum at 90 ml. The position of this eluent fraction corresponded to the ascending shoulder of the second peak. The peak in staining activity did not correspond to a peak in total protein. Thus, the bulk of product obtained from the conjugation reaction was not enzymatically and immunologically active conjugate. Even the use of purified antibodies for conjugation would not have averted these results.

The third peak consisted largely of PO, since none of the absorbance at 403 μ

under this peak was removed by 50% saturation with ammonium sulfate, conditions that leave PO in solution. The material under the first peak consisted of PO and antibody, but had no staining activity, and the PO absorbance was precipitated by 50% saturation with ammonium sulfate. Hence, this material represented large, inactive conjugates of PO with Ig or of Ig with Ig. The second peak represented unconjugated antibody. The elution pattern of the enzymatically and immunologically active conjugate suggests that it consisted of aggregates of various proportions of PO molecules with 1 or 2 molecules of IgG.

Preparation of Glucose Oxidase Antibody Conjugates with Glutaraldehyde

In the procedure recommended by Avrameas (1969), 0.05 ml of a 1% aqueous solution of glutaraldehyde was added dropwise under gentle stirring to 5 mg of purified antibody (Avrameas and Ternyck, 1969) and 10 mg glucose oxidase. After 2 hr at room temperature, the solution was dialyzed against PBS at 4°C overnight. A precipitate was removed in a Spinco Rotor 40 at 20,000 rpm at 4°C. The conjugate solution was stable at 4°C for at least three months. An analogous procedure was used for the preparation of conjugates of antibody with peroxidase, phosphatase, and tyrosinase.

Electrophoresis has shown that the negative charge of the conjugate was increased relative to that of the original proteins, indicating that conjugation with glutaraldehyde was extensive: a large proportion of free amino groups in the proteins had reacted with the aldehyde, and each protein molecule was coupled with several aldehyde molecules. Analytic sedimentation confirmed this behavior: no major component of the conjugate possessed the sedimentation coefficient of the original Ig used. Consequently, no significant proportion of antibody remained unconjugated. This contrasts with the conjugation procedure using FNPS, in which a large proportion of antibody remained unconjugated and had to be separated from the conjugate by gel filtration (as peak No. 2, described above) to avoid competition for tissue antigen with the conjugated antibody during the staining reaction.

After glutaraldehyde conjugation, no single sedimentation peak could be attributed to the conjugate. Indeed, the conjugate may have been so heterogenous that no single species was represented in sufficient quantity to give a discernible peak in the Schlieren analysis of change in protein concentration with distance in the ultracentrifuge cell. Thus, it is unlikely that a complex of one antibody molecule with one or two enzyme molecules was a major portion of the total conjugate. Much of the conjugate probably consisted of large aggregates containing several antibody molecules each. In the conjugates prepared with FNPS, on the other hand, the immunologically and enzymatically active conjugate consisted of monomers or dimers of immunoglobulin coupled to PO.

It has been claimed that conjugation with glutaraldehyde is a mild procedure

(Avrameas, 1969), since reaction with amino group does not abolish antibody (Pressman *et al.,* 1961) or antigen reactivity (Pierce *et al.,* 1964; Sternberger, 1967; Avrameas and Ternyck, 1969). This is borne out by the fact that cross-linking with glutaraldehyde, extensive as it is, does not abolish antibody and enzyme reactivity in every conjugate molecule obtained. For a general discussion on the ability of glutaraldehyde to cross-link proteins, including enzymes, the reader is referred to Hayat (1970).

During conjugation with glutaraldehyde, the pH is maintained at 6.8, while conjugation with FNPS is carried out at pH 10.0. Unfolding of proteins at pH 10.0 and steric hindrance in refolding of the conjugate molecules could adversely affect the activity of the conjugate. Steric hindrance may also affect the glutaraldehyde-obtained conjugate, but the effect would be less pronounced than upon conjugation with FNPS.

Immunocytochemical Staining with Enzyme-Conjugated Antibodies

Thus far the indirect method has been used nearly exclusively. Tissues are first reacted with unconjugated antibodies to the antigen to be localized. Following washing, the specifically attached Ig is reacted with enzyme-conjugated anti-Ig. Usually, Ig fractions, rather than isolated antibody, have been used for conjugation, and specific staining has been obtained with FNPS, glutaraldehyde, or carbodiimide-conjugated preparations. PO has been the most frequently used enzyme.

Direct methodology, as described below, was employed when the antigen to be localized was itself an Ig, such as in autoimmune lesions (Fukuyama *et al.,* 1970). Direct methodology was also employed when the antiserum available was abundant, such as in the localization of tetanus toxin bound in striated muscle (Zacks and Sheff, 1968).

For light microscopy, fixation differs little in principle from that employed in immunofluorescence (von Mayersbach, 1966). For instance, human biopsy material from patients with cutaneous, autoimmune lesions (Wolff and Schreiner, 1970) is sectioned at 5 μ in the cryostat. The sections are placed on a microscope slide, washed for 15 min with PBS, and incubated for 3 hr at 22°C with conjugate diluted 1:10 in 0.1 M phosphate buffer (pH 7.2). Again, after washing, the sections are stained for 30 min with 0.05% DAB and 0.01% hydrogen peroxide (Graham and Karnovsky, 1966) and washed for 10 min in PBS. Sections can be observed with or without postosmication. Osmication is needed to obtain permanent preparations.

Electron microscopic staining can be carried out either on the tissue block prior to embedding or directly on the ultrathin sections. For staining prior to embedding, it is necessary to insure that the subcellular spaces are accessible to the antibody reagents. Normally, cells are not permeable to antibodies, even if unconjugated (Booyse *et al.,* 1971b). However, once a cell has been frozen and thawed or fixed, subcellular spaces may be accessible, even though penetration is slow (Pierce *et al.*, 1964). For this reason, exposure to antibodies, and particularly to conjugated antibodies, must be prolonged. This practice has to be extended to the washing period following application of each reagent. When a cell has been penetrated by conjugate sluggishly, removal of unreacted conjugate by washing is also expected to be slow.

A satisfactory method for staining prior to embedding has been offered by Nakane (1970) in his study on the immunocytologic classification of anterior pituitary cell types with regard to the hormone they secrete. Pituitary glands were fixed in phosphate-buffered 4% paraformaldehyde for 8 hr, washed in PBS overnight, and impregnated with 10% dimethyl sulfoxide (DMSO) for 1 hr for protection against damage by freezing. Frozen sections, 20 to 30 μ thick, were reacted with the following reagents and washing solutions for 24 hr each: specific antihormone sera from rabbits, PBS, PO-labeled sheep antirabbit Ig and PBS. The sections were then fixed in phosphate-buffered 5% glutaraldehyde for 4 hr, washed overnight, and impregnated for 1 to 2 hr with 0.05% DAB in 0.05 M Tris buffer (pH 7.6) brought to 0.01% in hydrogen peroxide for the final 15 to 30 min. Following washing in water for 2 hr and osmication for 4 hr, the sections were dehydrated in alcohol and embedded in Epon.

Staining on the methacrylate-embedded sections was reported by Kawari and Nakane (1970). A fixative must be chosen that does not abolish all antigenic reactivity of the antigen studied. For pituitary hormones, fixation for 4 to 8 hr in 4% paraformaldehyde and picric acid buffered with phosphate (Zamboni and DeMartino, 1967) has proved satisfactory. Following dehydration with ethanol, tissues were embedded in prepolymerized methacrylate. Polymerization was carried out at 4°C with ultraviolet light. Thin sections on collodion and carbon-covered grids were etched in benzene- or xylene-saturated water. After extensive washing in saline, they were floated on droplets of rabbit antisera reacting specifically with rat luteinizing hormone (LH), growth hormone (GH), and prolactin, respectively.

Following washing in saline, the sections were floated on PO-conjugated sheep antirabbit Ig. Again after washing, the grids were stuck on one edge of an 11 × 22 glass cover slip by means of doubly coated pressure-sensitive tape, placed in a 5 ml syringe through which the substrate for peroxidase was pumped at a rate of ~ 25 ml per min.

Either 0.005% DAB or a 1/10 saturated solution of 4-Cl-1-naphthol was used as electron donor in the substrate containing 0.001% hydrogen peroxide and 0.005 M Tris (pH 7.6). The continuous flow was necessary to prevent nonspecific adherence of the insoluble reaction product to the section and grid. Serial sections of rat pituitary were stained for LH and GH, respectively, and thus served as mutual controls. The reaction product formed deposits on secretion granules. Occasionally, electron-opaque deposits were also seen associated with the endoplasmic reticulum, but only in the very cell which also had specifically stained secretion granules. The deposits were 100 to 300 Å in diameter when DAB and 300 to 500 Å in diameter when 4-Cl-1-naphthol was used as substrate. These deposits filled a large proportion of the space within the secretion granules.

The most significant application of the PO-conjugated antibody method thus far has been the work of Nakane (1970) and Baker (1970), resulting in the classification of cell types responsible for secretion of six pituitary hormones. This was accomplished by low- and high-power light microscopy and by electron microscopy. Two, or even three, different hormones could be identified with light microscopy in the same section with their respective specific antibodies and different chromogenic substrates for peroxidase. For example, GH cells were stained blue by application of anti-GH rabbit serum and peroxidase-conjugated antirabbit Ig and 4-Cl-1-naphthol. The antibodies were then dissociated from the section either by acid or by 5 M sodium or potassium iodide, leaving bound the insoluble blue reaction products. Immunohistochemical staining was repeated using rabbit anticorticotropin (anti-ACTH), PO-conjugated antirabbit Ig and DAB, and hydrogen peroxide, which stained the ACTH cells brown (Nakane, 1968).

Another aid to the success of these studies was the correlation of light-microscopic immunohistochemical staining with the classical, nonimmunologic stains for differentiation of pituitary cell types (Baker, 1970). GH and prolactin cells were identified by light-microscopic immunohistochemistry, destained, and restained with the Masson trichrome procedure (Baker *et al.*, 1969). All acidophils whose cytoplasm stained uniformly and with varying shades of red following Masson staining were growth hormone cells. All acidophils containing large fuchsinophilic granules were prolactin cells.

The applicability of PO staining to both light and electron microscopy has facilitated low-power identification of the distribution of a specific cell type within the pituitary followed by selection of favorable regions for electron-immunocytochemical delineation of subcellular structures (Nakane, 1970). As a result, four out of six pituitary hormones have been shown to be secreted each by its own specific cell type (GH, ACTH, TSH, and prolactin). To a large extent, the "one hormone—one cell" concept is thus confirmed (Herlant, 1964; McShan and Hartley, 1965; Hilderbrand *et al.*, 1957). However, one of the cell types that had FSH in its secretion granules (oval cells with large vacuoles in their entire cytoplasm) also possessed LH. Secretion granules were dispersed among the vacuoles. This cell type was distributed along the periphery of the gland. Other cell types possessing FSH were devoid of LH (angulated cells which possessed secretion granules near the plasma membrane), and were located in the center of the gland. Nakane suggests an intimate functional relationship between the LH and FSH cells.

The prolactin cell was characteristically cup-shaped, frequently surrounded gonadotropic cells, and showed central nuclei. The secretion granules were large, and their rough endoplasmic reticulum was well developed (Nakane, 1970). Administration of norethynodrel enlarged these cells. It also caused reduction in size of the enlarged gonadotropic cells (presumably LH cells) found in ovariectomized rats (Baker and Yu, 1970). The close anatomic relationship of prolactin and gonadotropic (LH and FSH cells) has been suggested as significant in view of such inverse functional behavior (Nakane, 1970).

The ACTH cell was characteristically stellate, and its secretion granules were peripherally arranged. Cell processes often extended to sinusoidal walls. Destaining experiments support the notion (Hess *et al.,* 1968) that the ACTH cell may be a subclass of basophils (Baker, 1970), thus confirming similar investigations with fluorescent antibodies (Marshall, 1951; Leznoff *et al.,* 1962).

TSH was found in a polygonal cell clustered at the center of the gland. The distribution of secretion granules tended toward the cell periphery, and large vacuoles devoid of hormone were dispersed throughout the cytoplasm (Nakane, 1970). Cytologic alterations in TSH cells generally paralleled the differences in contents of thyrotropic hormone associated with sex, thyroid deficiency, thyroxine administration, and rebound after cessation of treatment with propylthiouracil (Baker and Yu, 1971).

The GH cell was oval, and possessed a central nucleus. The cell was dispersed evenly in most of the gland, but was found sparsely near the intermediate lobe and the anterior ventral portion. The secretion granules were dispersed throughout the cytoplasm (Nakane, 1970).

Several studies compared PO and fluorescein-labeled antibodies. Such studies led to the conclusion that epithelial cells *in vitro* synthesize their own basement membrane antigens, and that epithelial membranes of mouse, rat, and man contain species-specific antigens that do not cross-react with each other as well as species cross-reacting antigens that also react with collagen (Pierce and Nakane, 1967).

Using light microscopy, Davey and Busch (1970) have reproduced with PO-conjugated antibody the linear deposits on the glomerular basement membrane which had been established by immunofluorescence as characteristic of autoimmune antibody binding (nephrotoxic nephritis). Similarly, the granular immunofluorescent deposits of antigen-antibody complex nephritis (Boss, 1967) were reproduced by PO-conjugated antibody as large, coalescent deposits. Again reproducing immunofluorescence observations of autoimmune disease, linear deposits of PO reaction products were obtained in skin biopsies of pemphigoid lesions and granular deposits in biopsies of lupus erythematosus (Wolff and Schreiner, 1970). The deposits were found in the intercellular spaces in cases of pemphigus, close to the basement membrane in pemphigoid and lupus erythematosus lesions, and in the nuclei in several lesions of systemic lupus erythematosus (Fukuyama *et al.,* 1970).

Electron microscopy demonstrated immunoglobulins in lupus erythematosus as confined to the dermal-epidermal junction without exhibiting specificity toward any anatomically definable subcellular structure. They were not only found in the basal lamina but also appeared to be deposited randomly in the ground substance (Schreiner and Wolff, 1970). Electron-microscopic studies of experimental immune complex nephritis, using phosphatase-conjugated antibodies (Laguens and Segal, 1969), demonstrated autologous globulins in the epithelial half of the glomerular capillary basement membrane, in the cytoplasm of the foot processes, and sometimes diffusely in the basement membrane of the proximal tubules and at other times only in otherwise structureless areas within the basement membrane.

Thick sections of brown algae were reacted with rabbit antiserum to alginic acid followed by PO or fluorescein-labeled antirabbit Ig (Vreeland, 1970). Fluorescent staining was specific. Staining of intracellular regions with PO-labeled antibody was inconclusive because of nonspecific reaction by the substrate alone and by the labeled antibody preparation in the absence of antialginic acid antiserum. PO-labeled antibody staining of extracellular regions of high antigen concentrations was specific.

Zeromsky et al. (1970) demonstrated autoimmune antibodies in ulcerative colitis by means of fluorescein and PO-conjugated sheep antihuman antibodies. The PO-conjugate was made with specifically purified antibody, and contained as much as 0.12 to 0.5 mg antibody per ml. The fluorescent antibody, on the other hand, was a serum Ig fraction containing unspecified amounts of specific antibody. Although this is expected to favor efficiency of the PO-conjugate, there was no pronounced difference in specific contrast with both methods. However, no comparison of sensitivity was made by dilution of the reagents.

Avrameas (1969) compared the number of IgG-containing spleen cells stained in parallel with purified fluorescent anti-IgG and PO-conjugated anti-IgG. Twice as many cells were stained with the PO-labeled antibody. Phosphatase and glucose oxidase-labeled antibody stained essentially the same number of cells as did fluorescent antibody. It was concluded from these experiments that enzyme-labeled antibodies are at least as sensitive as fluorescein-labeled antibodies (Avrameas, 1969). At least equal sensitivities were attributed to PO-conjugated and to fluorescein-conjugated antibodies by Ubertini et al. (1971) in their identification of reovirus in cultured liver cells.

The main advantage of the enzyme-conjugated antibody method over older procedures is the applicability of the same method both for light and electron microscopy. Other frequently quoted advantages in comparison with immunofluorescence include permanence of stained preparations, at least if they are postosmicated, and the use of visible instead of UV light microscopy. The frequently quoted advantage of being able to increase sensitivity at will by permitting increased buildup of enzyme reaction product cannot be considered valid as it has failed to result in a sensitivity higher than that of immunofluorescence.

In view of the generally accepted amplification afforded by the use of an enzymatic detector, one wonders why the sensitivity of the enzyme-labeled antibody method is in the same range as that of fluorescent antibodies and not several orders of magnitude higher. Even the use of purified antibody in preparation of the conjugate, recommended to be essential by Avrameas (1969), did not increase sensitivity dramatically beyond that of fluorescent antibody. This unexpected failure of attainment of the greatest potential advantage of antibody enzyme methods has therefore directed attention to possible problems in the antibody conjugation procedure proper.

The enzyme-conjugated antibody methods are an extension of the ferritin-conjugated antibody method in that the preparation of the conjugate requires cross-linking of two different protein molecules *via* a bifunctional reagent. Essentially this is a reaction between three molecules, two large protein molecules and one small molecule of the cross-linking agent. In order to control this reaction, it is properly conducted in two steps: a relatively rapid step of reaction of the first protein with excess of cross-linking reagent, and a relatively slower step in the reaction of the conjugated first protein with the native second protein.

Singer and Schick (1961) were careful to adhere to this principle when conjugating ferritin to Ig in two steps. In the first step, ferritin was reacted with an excess of toluene 2,4-diisocyanate. One of the isocyanate groups formed ureido bonds with the ferritin molecule, while the other group remained free. The available amino groups of the ferritin molecule were thereby exhaustively reacted in a reaction step that can be permitted to go to completion. Alteration of the ferritin molecule was of little consequence, since the immunoferritin method does not depend upon the functional integrity of ferritin as long as the contents of iron micelles remains unaffected.

The second reaction step was carried out following removal of unreacted diisocyanate from the conjugated ferritin. This step consisted merely of an admixture of the reacted ferritin with unreacted Ig, and is therefore the reaction between two large molecules only. This insures, first of all, that ferritin will not react directly with ferritin to form large unreactive ferritin-ferritin aggregates, and, second, that Ig will not react directly with Ig to form large Ig-Ig aggregates that would be hard to separate from ferritin-Ig and would interfere with its localization. The only aggregates that can form are aggregates of ferritin with one or more Ig molecules or cross-linked multiples of these aggregates. Another advantage of the two-step method is the absence of excess diisocyanate in the second step. The only source of destruction of antibody specificity is therefore steric hindrance by conjugation with ferritin when it occurs at or near the antibody specific combining site.

In contrast to the ferritin-conjugated antibody method, in which only the functional integrity of the antibody has to be maintained, the enzyme-conjugated methods require functional maintenance of both antibody and enzyme. If one were to react one of these protein components with excess bifunctional reagent prior to reaction of both components with each other, one might expect that the component reacted in the first stage could have been destroyed. It is therefore not surprising that antibody-enzyme conjugates have been prepared by a single-step procedure in which antibody, enzyme, and an excess bifunctional reagent are mixed.

Two kinds of reactions occur simultaneously. The smaller bifunctional reagent reacts with one of the protein molecules and, at the same time, conjugated protein reacts with other conjugated protein. The latter reaction takes place between two large protein molecules, and hence possesses a lower collision frequency than the reaction of bifunctional reagent with one protein molecule. Therefore an excess bifunctional reagent is required for efficient conjugation. If the amount of reagent is limited, the slower protein-protein reaction is largely abolished. Instead, the free groups of the bifunctional reagent attached to the first protein would react with other sites within the same protein molecule or with surrounding water prior to cross-linking with a second protein.

Use of excess bifunctional reagent insures that at least some reactive groups in the first protein molecule will cross-link with a second protein molecule. During this reaction, both antibody and enzyme are exposed to a high concentration of bifunctional reagent, to which, in the ferritin-conjugated antibody method, only the ferritin is exposed. It is known that in the ferritin-conjugated antibody method in general only a fraction of the original antibody is obtained as immunologically active conjugate (Borek and Silverstein, 1961; Vogt and Kopp, 1965a and b). With special precautions, this fraction can be kept relatively high (Birnbaum et al. 1970). No data on the degree of retention of immunologically or enzymatically active material in enzyme-antibody conjugates are available. However, the exposure of antibody to high concentrations of bifunctional reagent suggests that the antibody active fraction must be much lower than in the ferritin method.

The products of conjugation of enzyme and Ig (which usually contains a small proportion of specific antibody in a large pool of nonspecific Ig) are expected to be:

1. Active enzyme-active enzyme
2. Active enzyme-inactive enzyme
3. Active enzyme-active antibody
4. Active enzyme-inactive antibody
5. Active enzyme-nonspecific Ig
6. Inactive enzyme-inactive enzyme
7. Inactive enzyme-active antibody
8. Inactive enzyme-inactive antibody
9. Inactive enzyme-nonspecific Ig
10. Active antibody-active antibody
11. Active antibody-inactive antibody

12. Active antibody-nonspecific Ig
13. Inactive antibody-inactive antibody
14. Inactive antibody-nonspecific Ig
15. Nonspecific Ig-nonspecific Ig

In addition, trimers, quatrimers, and polymers of any of these molecular conjugates are formed. Finally, the crude conjugate contains unconjugated (monomeric) molecules.

Filtration of PO-Ig conjugates through Biogel 300 removes unconjugated PO and also unreacted Ig (Nakane and Pierce, 1967). Since the conjugate effective in staining elutes between the void volume and before the unconjugated Ig, it is likely that the bulk of collected conjugate fractions consists of dimers of (3)-(5) and (7)-(15) plus Ig molecules conjugated with two or more PO molecules but not of conjugate containing one PO and more than two Ig molecules. Unfortunately, the gel filtration as recommended (Nakane and Pierce, 1967) is often being omitted or substituted by half-saturation with ammonium sulfate (Davey and Busch, 1970; Schiff et al., 1970). This treatment will separate the conjugate from unreacted PO but will not separate it from unreacted antibody or polymeric conjugate. Unreacted antibody interferes with the localization of conjugated antibody, and the polymeric conjugate is likely to impair resolution in electron microscopy.

A careful study by Modesto and Pesce (1971) has shown that conjugates (1) and (2) and (6)–(9) are not formed in significant quantities, even if large amounts of FNPS are mixed with PO. PO was found to be fairly unreactive with FNPS, so that even extensive conjugation did not inactivate the enzyme and did not lead to reaction of more than 1.6 moles of FNPS per mole of PO. FNPS reacts preferentially with Ig, leading to its precipitation. In the mixed reaction of PO, IgG, and FNPS, the FNPS exhibited 66-fold preferential binding with IgG over that with PO, and only a small proportion of total conjugate represented combination of IgG with PO. Conjugates (10)–(15) formed the bulk of the reaction product.

Since specific antibody is usually only a small proportion of total Ig, and only component (3) of the above mixture of fifteen conjugate components is immunohistochemically useful, the use of purified antibody has been recommended for the preparation of enzyme conjugates (Sternberger, 1967; Avrameas, 1969). This eliminates components (5), (9), (12), (14), and (15) from the mixture of conjugates obtained. Among the remaining impurities, (4), (8), and (13) can be removed from the tissue by washing during the staining procedure. However, conjugate of active antibody with active or inactive antibody will interfere with the localization of immunologically and enzymatically active conjugate. No method has yet been designed for separation of these components from the conjugate mixture.

Apparently because of these factors, the theoretically expected high sensitivity of enzyme-labeled antibody methods has not been materialized. In order to take advantage of the possibility for high sensitivity afforded by enzymatic detectors, a new method has been sought which uses enzymes without reducing yield of product as a result of conjugation and without producing substances that interfere with specific localization. This subject is discussed below.

THE UNLABELED ANTIBODY ENZYME METHOD

Enzymes as detectors of the immunohistochemical reaction are in principle capable of providing amplified sensitivity. To take advantage of this possibility, the immunologic and enzymatic reagents should possess maximal reactivity. This goal can be achieved by avoiding immunologically reacting materials that cannot be detected histochemically and that block reaction of the detectable material. Blocking materials that contaminate enzyme-antibody conjugates are unconjugated antibody and antibody conjugated with antibody rather than with enzyme.

The many side reactions in conjugation further reduce specific reactivity of conjugate as they diminish the yield of immunologically and enzymatically active materials. Finally, the lack or paucity in horseradish peroxidase of free amine groups, the points of attack of most commonly used cross-linking reagents (Ornstein, 1966), seems to be additional deterrent to conjugation. In order to insure enzymatic reactivity of *all* available antibody, it was therefore felt that conjugation must be avoided entirely, and a method was developed that used only native antibodies for linking antigen to enzyme.

For detection of an antigenic determinant "X," tissue is first reacted with specific antiserum to X produced in a suitable species—for example, rabbit (Fig. 7–1, Step I). Washing removes nonspecific protein and the tissue is reacted with antirabbit IgG produced in another species—for example, sheep (Step II). One of the specific combining sites of the anti-IgG (terminal on the FAb portion) reacts as antibody with the already localized rabbit anti-X, which is the antigen. Predominantly, the reaction of anti-IgG is with the Fc portion of the rabbit anti-X. The sheep anti-IgG is used in excess, so that only one FAb reacts, while the other FAb remains free (Lafferty and Oerteliss, 1963), thus providing an unreacted combining site specific for rabbit IgG. A number of sheep antirabbit IgG molecules react with each molecule of rabbit anti-X.

In Step III, purified rabbit anti-PO is applied. It reacts as an antigen with the free FAb site of rabbit anti-IgG. Most of the reaction of the anti-PO is *via* antigenic determinants on its Fc region. Even those anti-PO molecules that react *via* antigenic determinants on their FAb portions will leave free always one and usually both specific combining sites for PO. In Step IV, PO is applied. The PO reacts as antigen with the specific combining sites of anti-PO (the termini of the FAb portions). The bound enzyme is stained with DAB, hydrogen peroxide, and osmium tetroxide.

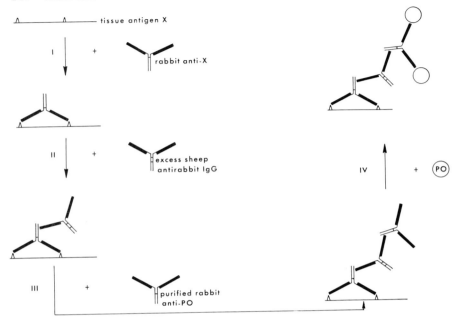

Fig. 7–1. Schematics of the unlabeled antibody enzyme method using purified anti-PO. Tissue antigen is localized by the sequential aμ μlication of antiserum produced in species A (I), antiserum to IgG of species A produced in species B (II), purified anti-PO from species A (III), and PO (IV) followed by staining with DAB and hydrogen peroxide and osmication (not shown). The heavy lines in the diagram of each IgG antibody molecule represent its two FAb portions, each containing one L-chain and the N-terminal part of one γ-chain. The two fine lines in each molecule represent the Fc portion consisting of the C-terminal parts of both γ-chains. In rabbit IgG antibody, the two γ-chains are bound by a single sulfhydryl linkage. In sheep IgG antibody, the γ-chains have been diagramed as being bound by two sulfhydryl linkages, but the actual number of inter-γ-chain linkages has not been established. By permission of Prentice-Hall, Inc., Englewood Cliffs, New Jersey. From the forthcoming *Immunocytochemistry* by Ludwig Sternberger.

When the above method was first reported (Sternberger, 1969), purified anti-PO was used in Step III. This was necessary because in antiserum or IgG obtained from rabbits immunized with PO, the IgG devoid of specificity for PO would react with the sheep anti-IgG equally well as would the anti-PO. This nonspecific rabbit IgG would interfere with localization of specific anti-PO. The sensitivity of the method would have been reduced, and the main rationale for which it had been developed would have been lost.

However, two months after acceptance of the first communication of this method, a paper was submitted elsewhere (Mason *et al.*, 1969), which described a method similar in all respects except for the use of antiserum to PO instead of purified antibody. Even with this simplification, specific localization was obtained, attesting to the sensitivity of the principle of using unconjugated anti-

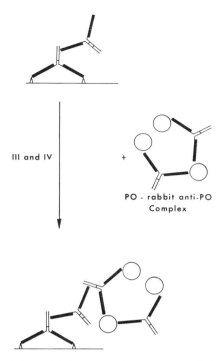

Fig. 7–2. Schematics of the unlabeled antibody enzyme method using peroxidase anti-peroxidase complex (PAP). Tissue antigen is localized as in Fig. 7–1, steps I and II. This is followed by application of PAP (III), DAB, and hydrogen peroxide and osmication (not shown).

bodies. However, when it is desired to localize every available antigen site at high resolution rather than only a minority of sites, purified antibody becomes essential. For the same reason, purified antibody seems to be necessary for the localization of sparsely distributed antigens.

In further developments of the method, the separate application of anti-PO (Step III) and PO (Step IV) was combined into a single step (Fig. 7–2). This was possible because antibody (bivalent) can combine with antigen (polyvalent) in multiple proportions, and complexes consisting of a high proportion of antigen to antibody are soluble (in contrast to complexes containing a high proportion of antibody to antigen, which form immune precipitates). The use of purified antigen-antibody complex consisting of PO and anti-PO replaced in a single step the sequential application of anti-PO and PO; the antibody moiety of the complex provided the antigenic determinants reacting with the sheep antirabbit IgG, and the PO moiety provided the immunohistochemically stainable enzyme.

The above simplification of the method was attempted for two reasons: First, excess free PO exhibits a degree of nonspecific binding for some tissue components (Behnke, 1968; White, 1970), while PO-anti-PO complex (PAP) does not (Booyse *et al.*, 1971b). Second, while preparation of purified anti-PO was a low yield procedure, a method had become available for preparing PAP at high yield (Sternberger *et al.*, 1970a).

Antibodies to protein antigens are commonly purified by the use of insolubilized antigens called immunoabsorbents. A large variety of procedures for insolubilizing protein antigens exist (Weliky and Weetall, 1965), including reactions with diazotized p-aminobenzyl cellulose (Campbell *et al.*, 1951), cyanogen bromide-treated Sepharose (Wide *et al.*, 1967; Ishizaka *et al.*, 1970; Mannik *et al.*, 1971), and glutaraldehyde (Avrameas, 1969). Antibodies are removed specifically from the solution by reaction with the immunoabsorbent. Following washings, the purified antibody is eluted from the immunoabsorbent, usually at acid pH. The yields vary among different antigens and among different antisera, but the average is $\sim 30\%$.

In general, only the fraction of antibodies with the weakest binding qualities is obtained, while the strongest binding antibodies remain attached to the immunoabsorbent (Schlossman and Kabat, 1962; De St. Groth, 1963; Kaplan and Kabat, 1966). In the case of anti-PO, dissociation from insolubilized PO at acid pH is exceedingly poor. This behavior resembles the resistance to acid dissociation of several other antigens and their specific antibodies—for example, p-aminohippurate (Bennett and Haber, 1963) or dinitrophenol (Weliky *et al.*, 1964)—and may perhaps be attributed to hydrophobic bonding.

Dissociation of antiperoxidase at pH 2.3 at 1°C from various immunoabsorbents (including glutaraldehyde-polymerized PO), although yielding only minimal amounts of purified antibody, was adequate for providing sufficient amounts of reagent for the unlabeled antibody enzyme method (Sternberger, 1969). However, this merely illustrates the sensitivity of the method. By using minimal amounts of anti-PO, the procedure is not employed in its optimal range of sensitivity. When anti-PO was dissociated from glutaraldehyde-aggregated PO at pH 2.3 at 25°C, large amounts of protein were eluted. This protein possessed a broad sucrose density sedimentation peak (7 to 8 s), and probably contained apo-PO bound to antibody (Hinton *et al.*, 1971). Only a small fraction of the total protein was free anti-PO antibody.

Purified antigen-antibody *complex* can be prepared by prolonged treatment of an immune precipitate with a fairly large excess of antigen, which causes slow resolubilization of the precipitate (Singer and Campbell, 1952, 1955; Ishizaka *et al.*, 1959; Sternberger *et al.*, 1961). The complexes so obtained are heterogeneous in antigen-antibody ratio, and contain a large amount of free antigen and denatured protein.

The following observation led to the development of a simple procedure for obtaining high yields of purified PAP using only moderate excess of antigen: Most protein-antiprotein immune precipitates readily dissolve at pH 2.3; PO-anti-PO immune precipitates do not. However, when PO-anti-PO precipitates are treated at pH 2.3 in the presence of a moderate excess of PO, resolubilization is rapid. It is assumed that at acid pH a rapid equilibrium is attained between a) PO-anti-PO precipitate; b) free PO and free anti-PO; and c) soluble PAP. The least stable form is (b). In the absence of added free PO, (a) will predominate, and the precipitate will fail to dissolve. In the presence of added free-PO, (c) will predominate, and soluble PAP can be obtained in high yields.

Preparation of Soluble Peroxidase-Antiperoxidase (Antigen-Antibody) Complex (PAP) (By permission of Prentice-Hall, Inc., Englewood Cliffs, New Jersey. From the forthcoming *Immunocytochemistry* by Ludwig Sternberger)

In a preliminary test, determine qualitatively the equivalence zone of the anti-PO serum used [prepared by immunization of rabbits with PO ($RZ^2 = 3.0$) and Freund's adjuvants]. Place into a series of ten tubes 0.2 ml saline containing amounts of PO varying from 0.05 to 0.5 mg per ml. Add to each tube 0.2 ml of anti-PO serum. Leave tubes at 1 to 5°C overnight, centrifuge, collect supernatants into a second set of tubes, and transfer 0.15 ml of each supernatant from the second set into a third set of tubes.

Add to each tube of the second set 0.1 ml of the anti-PO serum, and observe excess of antigen by qualitatively reading precipitation after 1 hr at room temperature. Add to each tube of the third set 0.1 ml of a solution of PO (0.1 mg per ml), and observe excess of antibody by qualitatively reading precipitation after 1 hr. Three zones will be observed: supernatants in the zone of excess antibody; supernatants in the equivalence zone, which show neither excess antibody nor excess antigen; and supernatants in the excess antigen zone. Record the concentration of PO *per* ml used in that tube of the equivalence zone which is nearest to the antigen excess zone (AgX equivalence proportion).

Good yields of PAP (in PO as well as in anti-PO) are obtained if anti-PO is precipitated from antiserum with 1.25 times AgX equivalence proportion. It is convenient to use 10 mg of PO for this initial precipitation. This determines the amount of antiserum required. If 10 mg is 1.25 times AgX equivalence proportion, then 8 mg is AgX equivalence proportion. Assume that in the preliminary test it was the tube to which 0.4 mg PO per ml (0.08 mg per tube) had been added to antiserum which precipitated at the AgX equivalence proportion. Then a total of 20 ml anti-PO serum will be needed to precipitate 8 mg PO at the AgX equivalence proportion or 10 mg at 1.25 times AgX equivalence proportion.

Therefore place in a 250 ml centrifuge bottle 4.0 ml of a 1:400 solution of PO freshly prepared by dissolving 63 mg of lyophylized PO (RZ = 3.0) in 25 ml of saline. Add 20 ml anti-PO serum and mix. Allow to stand at room temperature for 1 hr. Carry out the subsequent steps in refrigerated containers at 0 to 2°C or in a cold room at 0 to 5°C. Centrifuge at ∼ 2,000 rpm (International rotor 259) for 20 min. Remove supernatant by suction, and resuspend precipitate in a small volume of cold saline by forcing it several times through a 10 ml pipette. Wash by adding ~ 150 ml of saline and centrifuge. Carry out a total of three such washes.

Thoroughly resuspend the precipitate (by forcing it through a pipette) in 16 ml of the 1:400 solution of PO. Under mild stirring, bring to pH 2.3 at room temperature, with 1.0, 0.1, and 0.01 N HCl (1 drop of 1.0 N HCl followed by sufficient amounts of the more dilute solutions for fine adjustment). Neutralize immediately to ~ pH 7.4, using sodium hydroxide solutions (1.0, 0.1, and 0.01 N). This neutralization should be carried out irrespective of whether precipitate has dissolved completely or partially. Add 1.6 ml of a solution containing 0.08 N sodium acetate and 0.15 N ammonium acetate. Chill preparation in an ice bath. Carry out all subsequent steps in refrigerated containers at 0–2°C or in a cold room at 0–5°C. Centrifuge at ~ 17,500 rpm for 8 min (Sorvall rotor SS-34). Slowly add to the supernatant under stirring an equal volume of a solution of ammonium sulfate saturated at 0 to 5°C. Keep stirring for 25 min, and then centrifuge at 17,500 rpm for 16 min.

Wash the precipitate once in half-saturated ammonium sulfate solution. Dissolve the precipitate in 16 ml of water, and dialyze under protection from light against three changes of 15 liter each of sodium ammonium acetate saline (13.5 liter saline, 1.5 liter water, 75 ml of 1.5 N sodium acetate, and 75 ml of 3 N ammonium acetate solution). Centrifuge at 17,500 rpm for 16 min, quick-freeze 1 to 2 ml portions of the supernatant in dry ice and acetone, and store at −20°C in the dark. Alternatively, store the samples in a nitrogen freezer without preceding quick-freezing.

Prior to use, quick-thaw the samples of PAP and place them on an ice bath. Dilute the necessary amount of PAP as soon before use as possible, and keep on an ice bath. Quick-freeze the remaining undiluted PAP, and store for future use. Undiluted PAP stored in the frozen state for over eighteen months does not change in physical or staining properties. Dilute PAP, however, is unstable. When kept at 5°C, undiluted PAP has not been stable beyond a few weeks. The PO and anti-PO contents of PAP are determined by absorbance at 400 and 280 μ of samples diluted 1:5.

PO contents of PAP per ml = OD_{400} · 0.413 · 5 mg

Anti-PO contents of PAP per ml =

$$\left[OD_{280} - \frac{(OD_{400} \text{ of PAP}) \cdot OD_{280} \text{ of PO})}{(OD_{400} \text{ of PO})} \right] \cdot 0.620 \cdot 5 \text{ mg}$$

PO/anti-PO mole ratio = $\dfrac{mg\ PO \cdot 156{,}000}{mg\ anti\text{-}PO \cdot 39{,}800}$

PAP usually contains 1 to 2 mg of anti-PO per ml.

Properties of Peroxidase-Antiperoxidase Complex

Most soluble antigen-antibody complexes consist of heterogeneous mixtures of various ratios of antigen (Ag) to antibody (Ab), in addition to relatively large amounts of free Ag, such as $Ag_2 Ab$, $Ag_3 Ab_2$, $Ag_4 Ab_3$, . . . Ag_{n-1}, $Ag_n Ab_n$ . . . . Complexes on the right side of the series tend to precipitate. The change in free energy between one species of complex and another is small (Singer and Campbell, 1953; 1955). Only in the presence of a large excess of free antigen is the limiting complex of 2 Ag with 1 Ab molecule abundant. If free Ag is removed, complexes on the right side of the series predominate and precipitation will ensue.

When PAP was first prepared, two observations indicated that it was different from most soluble antigen-antibody complexes. In contrast to other soluble complexes, soluble PAP formed at high yields, with only moderate excesses of antigen [addition of an amount of PO four times that required for precipitation of anti-PO from antiserum when the composition of the precipitate was $(Ag_2 Ab_3)_n$]. Second, PAP separated from PO by ammonium sulfate precipitation remained soluble. Hence, it was suggested that PAP is an unusually stable molecular aggregate, and that the free energy of dissociation into other components is relatively large.

Further examination has shown that PAP is fairly homogeneous on boundary sedimentation and in the electron microscope. Unless the complex is zonally dispersed, it is maintained in solution in equilibrium with only a small amount of free antigen. Under zonal dispersion, such as gradient sedimentation or gel filtration, slow equilibration with free antigen and larger antigen-antibody aggregates ensues.

In a large series of preparations, the composition of PAP averaged 3 PO to 2 anti-PO molecules. If the molecular weight of PO is taken as 39,800 and that of anti-PO as 140,000 to 156,000, the ratio of PO to anti-PO suggests for PAP a molecular weight of 409,400 to 431,400 or multiples thereof. Sedimentation velocity measurement by Schlieren optics and sedimentation equilibrium measurements by UV optics independently arrived at molecular weights from 413,000 to 429,000—thus suggesting that PAP, indeed, consists of 3 subunits of PO and 2 subunits of anti-PO. This was confirmed by electron microscopy, which showed fair homogeneity after positive staining with DAB and hydrogen peroxide, followed by osmium tetroxide (Fig. 7–3) and after negative staining with phosphotungstic acid (Fig. 7–4). In both cases, PAP was a pentagonal molecular aggregate with average diameters of 205 Å.

After positive staining, the pentagons exhibited a central core; whereas after negative staining, the core was only occasionally apparent. It was concluded that corners 1 and 3 of the pentagon were contributed by the Fc fragments of anti-PO, and corners 2, 4, and 5 by PO (Fig. 7–5) (Sternberger *et al.*, 1970a). However, nothing was known regarding the forces that hold the 2 PO molecules together between corners 4 and 5.

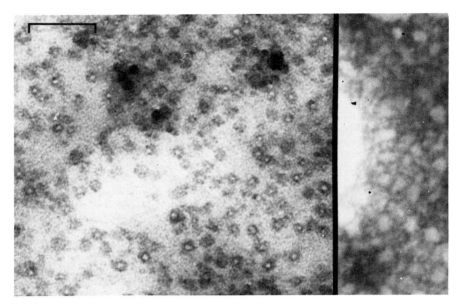

Fig. 7-3. (left) Peroxidase-antiperoxidase complex histochemically stained with DAB and hydrogen peroxide followed by osmication. Complex is ring-shaped, probably pentagonal, with average diameters of 205 Å. Marker is 1,000 Å.

Fig. 7-4. (right) Peroxidase-antiperoxidase complex negatively stained with phosphotungstic acid. Pentagonal molecules are seen. Magnification is the same as in Fig. 7-3.

Recent data (Hinton *et al.,* 1971) yielded information concerning these forces, and provided a possible clue for the circularity of PAP and its stability. It was found that PO polymerized at acid pH. Apo-PO (PO devoid of heme) did not polymerize. Conceivably, ring formation of PO and anti-PO depends upon capture of a dimeric PO molecule during acid polymerization of PO incumbent to the preparation of PAP. Ring formation was not observed when antigen-antibody complex was prepared by the acid treatment procedure, using apo-PO instead of PO.

Staining for Light Microscopy and Preembedding Staining for Electron Microscopy by the Unlabeled Antibody Enzyme Method

Fixation of tissues or cells is carried out with buffered paraformaldehyde or glutaraldehyde. The maximum fixation time and concentration of fixative is determined that leaves the reactivity of the particular antigen under investigation largely unaffected. For light microscopy, free cells or cryostat sections of the tissue are placed on glass slides covered with a thin layer of serum albumin or gelatin. Staining is carried out at room temperature with drops delivered from

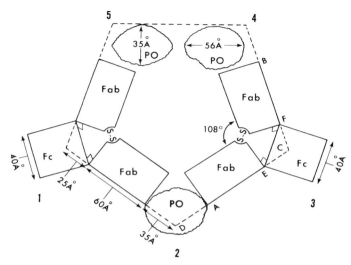

Fig. 7–5. Scale diagram of a PAP molecule. The calculated length of each side of the pentagon is 120 Å. Corners 1 and 3 are contributed by Fc of anti-PO and corners 2, 4, and 5 by PO. The ring is probably closed between the 2 PO molecules (4 and 5) by a polyheme bridge (Hinton *et al.*, 1971). From Sternberger *et al.*, 1970a.

Pasteur pipettes. The slides are kept in a moist chamber throughout the procedure. Suitable buffers for washing are Tris, phosphate, or veronal near neutrality.

The slides are treated in sequence with a 1:30 dilution of normal sheep serum in saline, varying dilutions of specific antiserum in buffer containing 0.1% gelatin, a 1:10 dilution of sheep antiserum to rabbit IgG in buffer containing 0.1% gelatin, and a 1:10 dilution of PAP in buffer containing 0.1% gelatin (the diluted PAP is kept in an ice bath). Between each step, the slides are rinsed with 10 ml of buffer delivered from a pipette, and then placed for 5 min in a jar containing the buffer. The area on the slides surrounding each spot of cells or sections is dried with filter paper after each wash to avoid spreading of drops of subsequently applied solutions.

Following application of PAP, the slides are washed in 0.05 M Tris buffer (pH 7.6) and stained for 5 min with a freshly prepared solution containing 0.05% DAB and 0.01% hydrogen peroxide in Tris buffer (Graham and Karnovsky, 1966). After being rinsed in water, the slides are placed in a chamber containing osmium tetroxide vapor (Seligman *et al.*, 1966) at 60°C for 30 min, or they are treated with a solution of buffered osmium tetroxide and washed in water (Caulfield, 1957).

For the localization of surface antigens on cells for electron microscopy, staining may be carried out in centrifuge tubes. The reagents are conveniently used in 0.2 ml amounts, and cells are exposed for 10 min to each protein solution and for 5 min to DAB and hydrogen peroxide. Cells are washed three times in ~ 3.0 ml of buffer between each step. Postosmication is carried out in buffered osmium tetroxide (Caulfield, 1957). The cells are dehydrated and embedded in Epon, Araldite, or low-viscosity Epon or Spurr mixture (Hayat, 1972).

For preembedding staining of subcellular antigens, fixed tissue is cut into 50 μ thick sections in a cryostat. The sections are stained with the same solutions as described above. However, even unlabeled antibodies penetrate sections of fixed tissue sluggishly (Pierce et al., 1964; Sternberger, 1967). Hence, both staining and washing must be carried out for extended durations. It is recommended that each reagent application step in the staining procedure, including the PAP step as well as the washings that follow, is carried out for ~ 24 hr, allowing ~ 8 to 12 hr for reagent application and the remainder of the duration for washing. Staining with DAB and hydrogen peroxide is carried out according to the procedure of Graham and Karnovsky (1966), and postosmication with buffered osmium tetroxide according to the procedure of Caulfield (1957). The sections are dehydrated and embedded according to standard procedures.

The unlabeled antibody enzyme method has been from one hundred to one thousand times more sensitive than immunofluorescence (Sternberger, 1971), as evidenced by the requirement of only one-one hundredth to one-one thousandth the concentration of antiserum for visualization of spirochetes in bright-field light microscopy. Since nonspecific staining of the organisms required equally high concentrations of normal serum in both methods, the unlabeled antibody enzyme method is also from one hundred to one thousand times more specific.

Kuettner et al. (1971) demonstrated by immunofluorescence and the unlabeled antibody peroxidase method that embryonic chick embryo cartilage is primarily extracellular in location. With immunofluorescence, specific localization was intense along the rim of the chondrocytes, corresponding to the anatomic lacunae. The unlabeled antibody enzyme method was more sensitive, since the intervening cartilage matrix was also stained, though at a lesser intensity. The high degree of sensitivity of the unlabeled antibody method was also demonstrated by the light-microscopic visualization of tetanus toxin in the perineurium of mice injected with minimal lethal amounts of tetanus toxin (Fedinik et al., 1970).

Additional antigens not previously detected by other immunocytochemical methods include the electron-microscopic demonstration of thrombosthenin on the surface of platelets (Booyse et al., 1971b). Thrombosthenin is the actomyosin-like protein involved in platelet aggregation and clot retraction. The existence of small amounts of surface thrombosthenin, in addition to thrombosthenin, inside the platelet had been surmised by the results of solubilization with high ionic strength sodium chloride solution (Booyse and Rafelson, 1971). With the unlabeled antibody enzyme method, it has been shown that, upon platelet aggregation by thrombin, thrombosthenin forms spicules on the surface of the cell. These spicules, upon progressive growth, become entangled in the platelet aggregate. The observations provide the first experimental evidence in favor of a contractile protein model for platelet aggregation (Booyse and Rafelson, 1971). The platelets of a patient with thrombosthenia were devoid of electron-microscopically detectable thrombosthenin (Booyse *et al.*, 1971a). When intact platelets were stained by the unlabeled antibody enzyme method, only surface thrombosthenin was revealed. The reagents did not penetrate the cell, even in the minimal amount detectable by the method.

The high sensitivity of the unlabeled antibody enzyme method is also illustrated by the fact that even when whole antiserum to PO is used instead of purified antibody, pituitary hormones are still localized at least as well as those obtained with the PO-labeled antibody method (Mason *et al.*, 1969). Similarly, chorionid gonadotropin was localized electron-microscopically in human term placenta with the use of whole anti-PO serum (Dreskin *et al.*, 1970). However, the same group of workers later reverted to the original method (Sternberger, 1969), using purified anti-PO in their studies on ACTH in the pituitary gland (Phifer *et al.*, 1970; Phifer and Spicer, 1970). Phifer (1970) also compared the sensitivity of the unlabeled antibody enzyme method with that of the enzyme-conjugated antibody method. Although only whole anti-PO serum was used in the former method, it was more sensitive than the latter method. A comparison of sensitivities of the unlabeled antibody enzyme method and the enzyme-conjugated antibody method during postembedding staining for electron microscopy is discussed below.

Two types of controls are indicated in the unlabeled antibody enzyme method. One is a method control, and consists of omission of the sheep antirabbit IgG serum. Nonspecific reaction of PAP, intrinsic tissue PO, and unusual osmiophilia of tissue are thereby controlled.

The other is a specificity control for evaluation of the degree of specificity of the antiserum to the tissue antigen used. The control consists of omission of the antiserum, substitution with normal serum, substitution with antiserum absorbed with specific antigen, and staining with specific antiserum of tissues known to be devoid of specific antigen. The high sensitivity of the unlabeled antibody enzyme method suggests that hitherto unsuspected antigen-antibody reactions become detectable. Indeed, even small amounts of antibodies present in normal sera may become reactive with many antigens under examination. Thus, undiluted normal sera nearly always reacted with spirochetes. These reactions can usually be detected only at high concentrations of normal serum, while reaction with specific antiserum is observed over a wide dilution range.

It has been possible to block the reaction of normal rabbit serum and spirochetes by pretreatment of the organisms with normal sheep serum, on the assumption that antibodies in normal sheep serum react with similar antigenic determinants of spirochetes as do antibodies in normal rabbit serum. Normal sheep serum is not detectable by the unlabeled antibody enzyme method, since sheep antirabbit IgG serum is unable to react with its own (sheep) IgG.

Tissues contain many components capable of nonspecific reaction with applied protein. A sensitive method may detect this reaction as false positive. The application of normal sheep serum prior to staining with rabbit antiserum and the unlabeled antibody enzyme method are in use, on the still unproved assumption that nonspecific absorption of normal sheep Ig by tissue would block nonspecific absorption of other Ig's subsequently applied. Thus, nonspecific absorption of rabbit IgG, of sheep IgG antirabbit IgG, and of PAP, which exhibits primarily an IgG surface, is prevented. However, nonspecific absorption of proteins other than IgG is not expected to be necessarily prevented. Free PO, for example, could still be absorbed nonspecifically. This is an important reason for using PAP instead of anti-PO followed by PO (Booyse *et al.*, 1971b).

Occasionally, the sheep antirabbit IgG serum may cross-react with normal IgG retained in the tissue to be stained. For this reason, it is necessary to preabsorb the sheep antiserum with normal serum of the species whose tissue is being studied. To 2.0 ml of sheep antirabbit IgG serum in a conical centrifuge tube, 0.2 ml of the respective normal serum is added. If after 30 min at 37°C a precipitate has developed, the mixture is centrifuged at 1°C at 2,400 rpm for 10 min (International rotor 253).

Without decanting the supernatant, another portion of 0.2 ml of normal serum is added, and the contents are mixed without disturbing the precipitate unduly. Incubation and centrifugation are repeated until no further precipitate develops after 30 min of incubation. A final portion of 0.2 ml of normal serum is added, and the mixture is left at 1 to 5°C overnight. Thus, if no precipitate develops after the first addition, at least two portions of normal serum have been added.

Postembedding Staining for Electron Microscopy by the Unlabeled Antibody Enzyme Method

When specific antiserum followed by sheep antirabbit IgG serum, PAP, and enzyme staining were applied in sequence to ultrathin sections prepared for electron microscopy, localization of electron-dense deposits was entirely nonspecific, and included even parts of the section devoid of tissue. Such nonspecific localization was not reported for the enzyme-conjugated antibody method (Kawari and Nakane, 1970). This was surprising, since the unlabeled antibody enzyme method uses no material which does not also enter into the reagents of the PO-conjugated antibody method.

Closer examination has shown that each of the protein reagents used in the unlabeled antibody enzyme method—namely, unlabeled rabbit antiserum, unlabeled sheep antirabbit IgG serum, and PAP—attached to embedding medium nonspecifically. It was also found that when any of these reagents is applied to the section first, it will block nonspecific absorption of a subsequent reagent. This provided a clue for specific staining on the ultrathin section. When the section was pretreated with normal sheep serum, none of the subsequently applied reagents of the unlabeled antibody enzyme method attached nonspecifically. The normal sheep serum itself is unreactive with any of these reagents (Hardy et al., 1970).

Pneumococcal polysaccharide has been localized on pneumococci embedded in cross-linked methacrylate, Epon, or in low-viscosity Epon (Hardy et al., 1970). Methacrylate sections were etched by flotation for 10 min on drops of water containing 0.01% benzene and 1% ethanol. Epoxy sections were etched by flotation for 10 min on drops of water containing 0.01% benzene, 5% methanol, and 5% ethanol. The drops had been placed on glass slides.

After etching, the grids were washed with a jet of water from a plastic spray bottle and placed on drops of the following solutions held in shallow depressions in a paraffin dish in a moist chamber: water for 30 min, normal sheep serum diluted 1:30 in saline for 5 min, varying dilutions of antiserum to pneumococcal polysaccharide in Tris gelatin for 5 min (0.05 M Tris buffer, pH 7.6, containing 0.15 M sodium chloride and 0.1% gelatin), sheep antirabbit IgG serum diluted 1:10 in Tris gelatin for 5 min, and PAP diluted 1:10 in Tris gelatin for 5 min. After each step, the grids were washed with a jet of saline from a plastic spray bottle and dried by touching them edgewise with a piece of filter paper.

After the last wash, the grids were placed for 2 min into a serum of 0.05 M Tris buffer (devoid of gelatin), pH 7.6, containing 0.0125% DAB and 0.0025% hydrogen peroxide delivered from a peristaltic pump at a rate of 6 to 20 ml per min. The grids were then washed in a jet of water, dried, and exposed to osmium tetroxide vapors (Seligman et al., 1966) at 47°C for 30 min. Nickel or gold grids were necessary because of the exposure to osmium tetroxide vapors. A similar procedure was used for the detection of single-stranded DNA in the heads of Epon-embedded λ-phage. Intraplatelet thrombosthenin has also been revealed on the methacrylate sections (Booyse et al., 1971b).

The ACTH-secreting cell has been examined by Moriarty and Halmi (1972) using postembedding staining with diluted rabbit antiserum against the synthetic peptide comprising amino acids 17–39 of ACTH, followed by goat antirabbit IgG serum and PAP. After treatment with DAB and hydrogen peroxide, osmication of the copper grids was carried out by flotation on buffered osmium tetroxide for 30 min. The secretion granules of the ACTH cells were sharply delineated (Figs. 7–6, 7–7, 7–8). At times the cytoplasm immediately surrounding the secretion granules also contained ACTH, as indicated by the specific deposits of the pentagonal PAP molecules (Fig. 7–7).

Since each antigen site is represented on the average by a number of PAP molecules (Sternberger et al., 1970a), the discrete localization of PAP outside

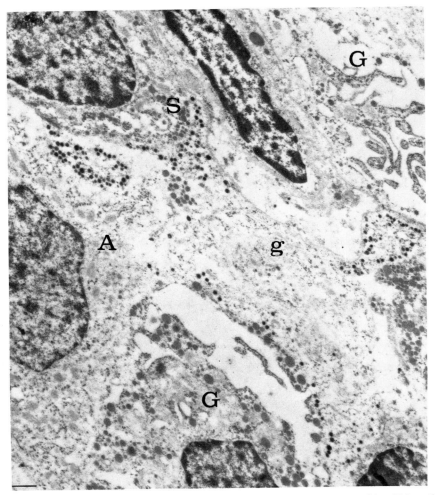

Fig. 7–6. Rat pituitary, Methacrylate-embedded. Stained on the section with rabbit anti-ACTH peptide 17-39 absorbed with α-melanocyte-stimulating hormone (2 mg/ml), diluted 1:100 and the unlabeled antibody enzyme method. Counterstained with lead citrate and uranyl acetate. The peripherally arranged granules of the ACTH cell (A) are opaque. g, Golgi. S-somadotrop. F, gonadotrop. Line at left lower corner, 1 μ. From the work of Gwen Moriarty and N. S. Halmi. For superior tissue preservation after staining on Araldite sections, and for recent methodologic advances see Moriarty and Halmi, 1972.

the secretion granules indicates a relatively sparse distribution of ACTH distinct from its high concentration within the secretion granules.

Moriarty and Halmi (1972) also carried out a comparison of sensitivities of the unlabeled antibody enzyme method and the PO-conjugated antibody method, using in postembedding staining the same antiserum to the 17–39

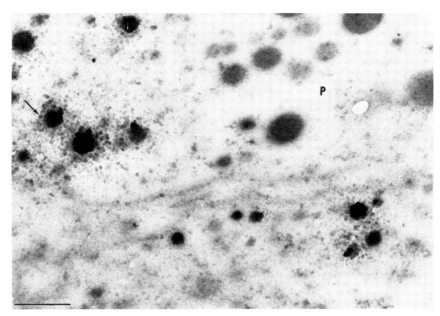

Fig. 7–7. Rat pituitary, Methacrylate-embedded. Stained on the section with rabbit anti-ACTH peptide 17-39, diluted 1:20, and the unlabeled antibody enzyme method. No counterstain. The diameter of the opaque granules in the ACTH cell processes is about 300 μ. Discrete PAP molecules (arrow) are in the cytoplasm surrounding the granules. P, prolactin cell granules. Line at left lower corner, 0.5 μ. From the work of Gwen Moriarty and N. S. Halmi.

peptide of ACTH for both methods. At a 1:500 dilution of anti-17−39, the unlabeled antibody enzyme procedure still yielded definite staining in the granule center. With PO-conjugated antibody method, there was little staining at a 1:100 dilution of the antiserum.

The localization of pentagonal structures as observed by Moriarty (Fig. 7−7) in postembedding staining can be used to differentiate the specific immunohisto-chemical reaction from the presence of intrinsic peroxidase. The pentagonal structure is not preserved, however, after preembedding staining.

Selective Suppression of an Immunohistochemical Reaction

Suppose one would like to use an antibody produced in species X to localize an antigen in a cell from the same species. It is assumed that the cell contains autologous Ig. Both the autologous Ig and the antibody applied to localize the antigen in question would react with the anti-X IgG used in the unlabeled antibody enzyme method. A procedure for selective suppression of reaction of autologous IgG is therefore desirable. Monovalent fragments (FAb) of IgG provide this capability.

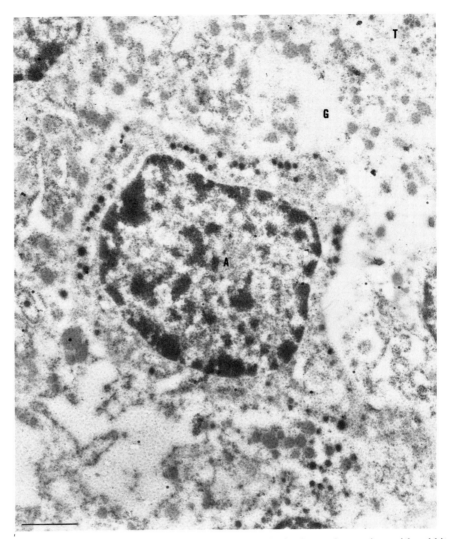

Fig. 7–8. Rat pituitary, Methacrylate-embedded. Stained on the section with rabbit anti-ACTH peptide 17-39, diluted 1:500, and the unlabeled antibody enzyme method. No counterstain. Even at this high dilution, the peripherally arranged granules of the ACTH cell (A) are opaque. G, gonadotroph. T, thyrotroph. Line at left lower corner, 1 μ. From the work of Gwen Moriarty and N. S. Halmi.

In a classical experiment, Porter (1959) showed that mild digestion of IgG with papain in the presence of cysteine splits IgG into crystalline fragment Fc and two heterogeneous fragments, FAb. Each FAb fragment bears one of the two combining sites of the antibody. Because of its monovalence, FAb does not precipitate with polyvalent antigen, and, indeed, efficiently inhibits the precipitation of undigested antibody and antigen.

In the unlabeled antibody enzyme method, substitution of sheep antirabbit IgG FAb for sheep antirabbit IgG serum has prevented the cross-linking of antitissue antibody with PAP. FAb can therefore be used to permit one anti-IgG reaction to be detected to the exclusion of another. This approach was used in the localization of C3 on sensitized sheep erythrocytes (SRBC). The cells had been reacted with rabbit anti-SRBC serum and with complement components 1, 4, 2, and 3 (Mayer, 1970). C3 was localized by the use of rabbit anti-C3 serum (courtesy of Dr. H. Shin).

In order to limit reaction of subsequently applied sheep antirabbit IgG to the anti-C3, and to exclude reaction with the anti-SRBC, sheep antirabbit IgG FAb was applied prior to anti-C3 and the unlabeled antibody enzyme method (Sternberger, 1972). C3 was localized over the entire surface of the SRBC. No staining of C3 occurred when the anti-C3 had been absorbed with C3, showing that the histochemical reaction with anti-SRBC had been eliminated. In separate staining experiments, the anti-SRBC was localized in discrete spots on the SRBC surface.

Approach to Quantitative Staining Histochemistry

Histochemistry has been largely a qualitative science. Unlike other biochemical reactions, histochemical reactions *in situ* cannot be expressed in terms of molar concentrations. The use of molarity for quantification depends upon the fact that molecules in solution possess a statistically random distribution. In order to quantitate biochemical reactions of tissues, it is therefore necessary to isolate a specific tissue component and measure it under conditions in which its molecular distribution, at least at the time of measurement, is uniform and random. Staining histochemistry, by requiring the examination of tissue constituents *in situ,* permits such randomization only at low resolutions, at which the assumption that distribution is not uniform is inconsequential. At high resolutions, however, such randomization is precluded.

Indeed, during the development of new cytochemical methods for high resolution, special care must be taken to avoid dislocation and randomization of tissue constituents. Since the point concentration of tissue constituents varies from one microscopic area to another, quantification in terms of molar chemistry is not possible. As pointed out by Glick (1967), one can contrast histochemical staining and quantitative chemical methods.

For this reason, it is desirable to use histochemical microscopy in combination with quantitative biochemical methods. Histochemistry benefits from being considered in a broad sense (Barka, 1971), which includes, among others, physical separation of tissue components. Extensive studies by Glick (1961; 1963) have contributed significantly to the advancement of quantitative histochemistry. In these studies, qualitative microscopic examination was correlated with careful quantitative measurements on defined amounts of tissue in terms of moles or moles per minute. This approach is correlative in the sense that quantitative biochemistry is related to separate anatomic observation.

A correlative approach may ultimately yield a direct quantitative histochemistry by narrowing the correlation to ever decreasing samples of the tissue: an area may be observed in the microscope and then analyzed *in situ.* Examples of such approaches are: vaporization, with a laser beam, of constituents of a sampling spot observed with light microscopy followed by spectrum analysis of the vapor (Glick, 1966); integrated scanning spectroscopy of the tissue (Carpenter, 1966); and absorbance measurement of individual electron-microscopically localized spots (Sternberger *et al.,* 1970a).

These methods depend upon simultaneous recording of the physical effects of a number of molecules on a detector, and therefore cannot be applied to the ordered arrangement of single molecules morphologically detectable in an electron microscope. To provide information on the quantitative distribution of chemical constituents of the cell at high resolution, enumeration of single molecules, once they are chemically marked, seems to be essential. Results would have to be expressed in terms of molecules per defined structure rather than moles observed in or removed from a defined area.

Requirements for such a method are the assurance that each histochemically positive spot represents reaction of not more than one molecular site *in situ.* Quantification is accomplished by comparing the number of positive molecules among morphologically different structures within single cells and among morphologically similar structures within different cells. Quantification in absolute terms is carried out by measuring the total number of positive molecules in a volume or weight-defined amount of the tissue under examination.

When an electron-immunocytochemical reaction is carried out under suboptimal conditions, such as when the concentration of antibody is progressively decreased, two events can occur: either the opacity of specific localization decreases, or the opacity remains the same but the number of sites specifically localized decreases. Only if the sensitivity and resolution of the method are high will the latter happen.

If with increasing dilutions the number of sites localized decreases while the intensity per localized site remains the same, the probability increases that each site represents an antigenic determinant within a single molecule. If one subcellular structure (e.g., a lysosome) contains one hundred times as many such sites as another structure (e.g., a unit length of rough endoplasmic reticulum), it is likely that the lysosome contains one hundred times as much of the structure as the length of endoplasmic reticulum (postembedding staining is required, or uniformity of penetration during preembedding staining must be assured, or quantification must be limited to cell surface antigens).

Studies on the reaction of SRBC with rabbit antibodies (Sternberger *et al.,* 1971) show that quantification by this approach is, indeed, feasible (Figs. 7–9,

7–10, 7–11, 7–12, 7–13). The rabbit antiserum to SRBC contained three subclasses of IgG, negligible amounts of IgM, and small amounts of an unidentified immunoglobulin class with anti-SRBC activity. The agglutination titer was 1:1,000—that is, at this dilution, a detectable amount of SRBC became agglutinized, while the bulk of SRBC remained single. Staining with this concentration of antiserum and the unlabeled antibody enzyme method showed that much of the surface of the erythrocyte was covered with antibody (Fig. 11). Confluence of points was extensive. It was impossible to enumerate individual sites, since single sites in all probability represented a number of anti-SRBC molecules.

The very minimal agglutination of cells, despite their extensive covering with anti-SRBC, seems to indicate that most antibodies utilize both of their combining sites for reaction on a single cell, and only a few cross-link two cells. At a concentration of anti-SRBC of 1:100 nearly all the surface (Fig. 7–10) and at 1:10 the entire surface of the cell (Fig. 7–9) was covered, showing that antigenic determinants are distributed throughout the surface of the SRBC. Upon dilution

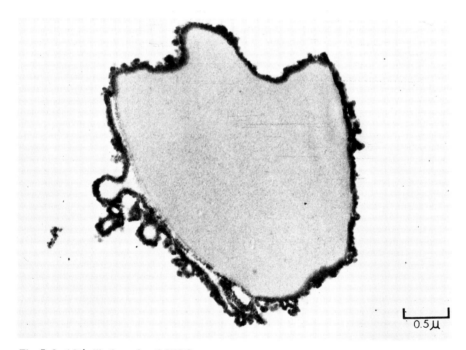

0.5μ

Fig. 7–9. 10^{-1} dilution of anti-SRBC.

Fig. 7–9–7–13. Sheep erythrocytes (SRBC) stained with varying dilutions of rabbit-anti-SRBC and the unlabeled antibody enzyme method. By permission of Prentice-Hall, Inc., Englewood Cliffs, New Jersey. From the forthcoming *Immunocytochemistry* by Ludwig Sternberger.

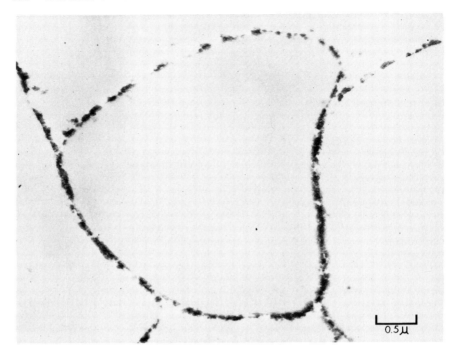

Fig. 7–10. 10^{-2} dilution of anti-SRBC.

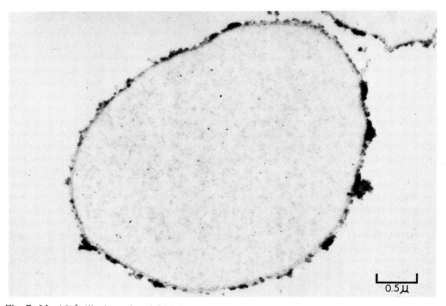

Fig. 7–11. 10^{-3} dilution of anti-SRBC.

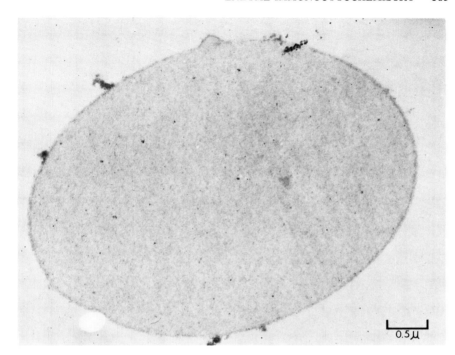

Fig. 7–12. 10^{-4} dilution of anti-SRBC.

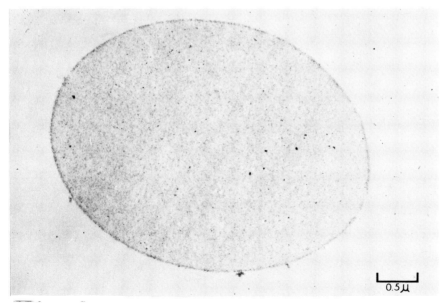

Fig. 7–13. 10^{-5} dilution of anti-SRBC.

of anti-SRBC to 1:10,000, antigenic sites became discrete (Fig. 7—12). Finally, at a concentration of 1:100,000, or one-one hundredth of the limiting agglutinating concentration, only a few sections contained a localized antigen site (Fig. 7—13).

At high dilutions, most of the antigen-bearing sites remained invisible, but those visualized were seen at an opacity equal to that of a point of equal size within the continuously localized area of staining after application of high concentrations of anti-SRBC. Cells prepared after application of anti-SRBC diluted 1:10 and the unlabeled antibody peroxidase method in which anti-RIgG was omitted, or cells prepared by the unlabeled antibody peroxidase method after omission of anti-SRBC, never provided an opaque site.

The number of opaque sites per section increased proportionally by increasing the concentration from 10^5 to 10^4, but evaluation of the total number of sites became impossible because of rapid confluence of spots with higher concentrations. Hence, the use of high dilutions of antiserum only helps to determine the relative number of antigenic sites among various structures in the same cell or tissue preparation. To determine the total number of antigen sites, independent measurements are needed. To accomplish this, a sensitive method for estimation of PO activity was developed.

Measurement of PO activity of cell suspensions after application of anti-SRBC at a 10^{-4} dilution, anti-RIgG, and PAP gave an estimate of the specific activity per single antigen site. The increase of peroxidase activity with increase of anti-SRBC was typical of a saturation curve. The saturation, in all likelihood, is due not only to exhaustion of available binding sites for antibody on the erythrocyte but also to steric hindrance of PAP when crowded on the erythrocyte surface. There may also be a steric effect due to crowding of anti-RIgG, but this will be less pronounced than the crowding of PAP, because of the smaller size and greater flexibility of the former.

In this section, an approach to quantitative staining histochemistry has been outlined, although no absolute measurements of total sites have yet been accomplished. The above discussion was presented because of a belief that in quantification lies the greatest hope for histochemistry. This is examplified by the evaluation of cellular cooperation in the immune response (Brody, 1970; Paul et al., 1970). At present, the most successful method for evaluating a completed immune response on the cellular level is by plaque formation as a result of antibodies secreted by cultured cells diluted sufficiently for one plaque to represent one cell (Jerne et al., 1963). Thus, the number of cells that have gone through a certain stage of the immune response at the time of plating can be evaluated.

Measurements of cellular cooperation depend upon bringing the final antibody-producing cell through the stage at which it can, after culturing, secrete the minimal amount of detectable antibody. Differences among individual cells remain largely unmeasured. With a quantitative histochemical method that enumerates the number of antibody molecules per unit of cell surface or cell section, one may perhaps differentiate several classes of cells at any instance of an immune response. Thus, more direct information can be obtained on individual stages in the cooperation of cells that yield an antibody-producing cell. Ficoll or serum albumin gradients (Gorczynski *et al.*, 1970; Aisenberg, 1971) may be used to separate cells that possess different amounts of antibodies or antigen-recognition sites on their surfaces, although eventually on culture each cell may produce a single plaque.

During the past decade, much biochemical experimentation has relied upon the relationship between soluble and particulate fractions. Perhaps the present decade will see growing concern with cell membranes. With such development, direct quantitative observation of organized structures may become increasingly important, and may place increasing demands on cytochemistry, especially quantitative cytochemistry.

For further discussion and viewpoints on enzyme immunocytochemistry see L. A. Sternberger, *Immunocytochemistry,* Prentice-Hall, Inc., New Jersey, in press.

The abbreviations used are: ACTH, corticotropin; C3, third component of complement; DAB, 3,3'-diaminobenzidine tetrahydrochloride; DMSO, dimethylsulfoxide; ER, endoplasmic reticulum; FAb, antibody fragment Ab; Fc, antibody fragment c; FNPS, 4,4'-difluoro-3,3'-dinitrodiphenyl sulfone; FSH, follicle stimulating hormone; GH, growth hormone; Ig, immunoglobulin; IgG, immunoglobulin G; IgM, immunoglobulin M; LH, luteinizing hormone; OD, optical density; PAP, soluble peroxidase antiperoxidase (antigen-antibody) complex; PBS, phosphate-buffered saline; PL, prolactin; PO, horseradish peroxidase; RIg, rabbit Ig; RIgG, rabbit IgG; SRBC, sheep red blood cells; TSH, thyrotropic hormone.

REFERENCES

Aisenberg, A. C. (1971). Size density analysis of rodent lymphoid cells. *J. Immunol.* **107**, 284.

Avrameas, S. (1969). Coupling of enzymes to proteins with glutaraldehyde: Use of the conjugates for the detection of antigens and antibodies. *Immunochem.* **6**, 43.

Avrameas, S., and Lespinats, G. (1967). Detection of intracellular antibodies in immunocompetent cells of animals immunized with enzymes. *C. R. Acad. Sci.* (Paris), Ser. D **265**, 302.

Avrameas, S., and Ternyck, T. (1969). The crosslinking of proteins with glutaraldehyde and its use for the preparation of immunoabsorbents. *Immunochem.* **6**, 53.

Avrameas, S., and Uriel, J. (1966). Methods for labeling antibodies and antigens with enzymes: Application in immunodiffusion. *C. R. Acad. Sci.* (Paris), Ser. D **262**, *2543*.

Baker, B. L. (1970). Studies on hormone localization with emphasis on hypothesis. *J. Histochem. Cytochem.* **18**, 1.

Baker, B. L., Midgley, Jr., A. R., Gersten, B. E., and Yu, Y. Y. (1969). Differentiation of growth hormone and prolactin-containing acidophils with peroxidase-labeled antibody. *Anat. Rec.* **164**, 163.

Baker, B. L., and Yu, Y. Y. (1970). The influence of norethynodrel on the hypophysis. *Proc. Soc. Exp. Bio. Med.* **134**, 107.

Baker, B. L., and Yu, Y. Y. (1971). The thyrotropic cell of the rat hypophysis as studied with peroxidase-labeled antibody. *Anat. Rec.* **169**, 270.

Barka, T. (1971). Editorial policy and practices. *J. Histochem. Cytochem.* **19**, 201.

Behnke, O. (1968). Electron microscopical observations on the surface coating of human blood platelets. *J. Ultrastruct. Res.* **24**, 51.

Bennett, J. C., and Haber, E. (1963). Studies on antigen confirmation during antibody purification. *J. Biol. Chem.* **238**, 1362.

Birnbaum, U., Vogt, A., and Marinis, S. (1970). Isolation and characterization of immunoferritin conjugates. II. Antibody binding capacity *in vitro. Immunol.* **18**, 443.

Booyse, F. M., Kisieleski, D., Seeler, R., and Rafelson, Jr., M. E. (1971a). Thrombosthenin defect in Glanzmann's thrombasthenia. Blood (in press).

Booyse, F. M., and Rafelson, Jr., M. E. (1971). Human platelets contractile proteins: location properties and function. *Ser. Haematolgica* **4**, 152.

Booyse, F. M., Sternberger, L. A., Zschocke, D., and Rafelson, Jr., M. E. (1971b). Ultrastructural localization of contractile protein (thrombosthenin) in human platelets using an unlabeled antibody peroxidase staining technique. *J. Histochem. Cytochem.* **19**, 540.

Borek, F., and Silverstein, A. M. (1961). Characterization and purification of ferritin-antibody globulin conjugates. *J. Immunol.* **87**, 555.

Boss, J. H. (1967). Anti-kidney antibodies. *Israel J. Med. Sci.* **3**, 167.

Brody, T. (1970). Identification of two cell populations required for mouse immunocompetence. *J. Immunol.* **105**, 126.

Campbell, O. N., Luescher, I., and Lerman, L. S. (1951). Immunologic absorbents. I. Isolation of antibody by means of a cellulose-protein antigen. *Proc. Nat. Acad. Sci.* **37**, 575.

Carpenter, A. (1966). Scanning methods: volume quantitation of tissues, cells, and subcellular components. *J. Histochem. Cytochem.* **14**, 834.

Caulfield, J. B. (1957). Effect of varying the vehicle for osmium tetroxide in tissue fixation. *J. Biophys. Biochem. Cytol.* **3**, 827.

Coombs, R. R. A., Mourant, A. E., and Race, R. R. (1945). A new test for the detection of weak and "incomplete" Rh agglutinins. *Brit. J. Exp. Path.* **26**, 255.

Coons, A. H. (1961). The beginnings of immunofluorescence. *J. Immunol.* **87**, 499.

Coons, A. H., and Kaplan, M. H. (1950). Localization of antigen in tissues: improvement in a method for the detection of antigen by means of fluorescent antibody. *J. Exp. Med.* **91**, 1.

Crowle, A. J. (1961). *Immunodiffusion.* Academic Press, New York.

Davey, F. R., and Busch, G. J. (1970). Immunohistochemistry of glomerulonephritis using horseradish-peroxidase and fluorescein-labeled antibody: a comparison of two techniques. *Amer. J. Clin. Pathol.* **53**, 531.

De St. Groth, S. F. (1963). Discussion on specific site topology of enzymes and antibodies induced by the same determinants. *Ann. N.Y. Acad. Sci.* **103**, 609.

Dreskin, R. B., Spicer, S. S., and Greene, W. B. (1970). Ultrastructural localization of chronic gonadotropin in human term placenta. *J. Histochem. Cytochem.* 18, 862.

Fedinek, A. F., Deboo, M. M., Gardner, D. P., and Sternberger, L. A. (1970). Immunohistochemical localization of tetanus toxin by the unlabeled antibody peroxidase method. *J. Histochem. Cytochem.* 18, 684.

Fife, Jr., E. H., and Muschel, L. H. (1959). Fluorescent antibody technique for the serum diagnosis of *Trypanosoma cruzi* infection. *Proc. Soc. Exp. Biol. Med.* 101, 540.

Franz, H. (1968). Basis for a new method of electron microscopic immunohistochemistry. *Histochemie* 12, 230.

Fukuyama, K., Douglass, S. D., Tuffanelli, D. L., and Epstein, W. L. (1970). Immunohistochemical method for localization of antibodies in cutaneous disease. *Amer. J. Clin. Pathol.* 54, 410.

Glick, D. (1961). *Quantitative Chemical Techniques of Histo- and Cytochemistry,* Vol. 1. Wiley-Interscience, New York.

Glick, D. (1963). *Quantitative Chemical Techniques of Histo- and Cytochemistry,* Vol. II. Wiley-Interscience, New York.

Glick, D. (1966). The laser microprobe: Its use for elemental analysis in histochemistry. *J. Histochem. Cytochem.* 14, 862.

Glick, D. (1967). Usage of "histochemical," "staining," and "biochemical" in histochemical literature. *J. Histochem. Cytochem.* 15, 299.

Goldfischer, S., and Essner, E. (1970). Peroxidase activity in peroxisomes (microbodies) of acatalasemic mice. *J. Histochem. Cytochem.* 18, 482.

Goldstein, G., Slizys, I. S., and Chase, M. W. (1961). Studies on fluorescent antibody staining. I. Non-specific fluorescence with fluorescein-coupled sheep antirabbit globulins. *J. Exp. Med.* 114, 89.

Gorczynski, R. M., Miller, R. G., and Phillips, R. A. (1970). Homogeneity of antibody-producing cells as analyzed by their buoyant density in gradients of Ficoll. *Immunol.* 19, 817.

Graham, Jr., R. C., and Karnovsky, M. J. (1966). The early stages of absorption of injected horseradish peroxidase in the proximal tubules of mouse kidney: ultrastructural cytochemistry by a new technique. *J. Histochem. Cytochem.* 14, 291.

Hardy, Jr., P. H., Meyer, G. H., Cuculis, J. J., Petrali, J. P., and Sternberger, L. A. (1970). Postembedding staining for electron microscopy by the unlabeled antibody peroxidase method. *J. Histochem. Cytochem.* 18, 678.

Hayat, M. A. (1970). *Principles and Techniques of Electron Microscopy: Biological Applications,* Vol. 1. Van Nostrand Reinhold Company, New York.

Hayat, M. A. (1972). *Basic Electron Microscopy Techniques.* Van Nostrand Reinhold Company, New York.

Henle, G. and Henle, W. (1965). Cross reactions among γ-globulins of various species in indirect immunofluorescence. *J. Immunol.* 95, 118.

Herlant, M. (1964). The cells of the adenohypophysis and their functional significance. *Int. Rev. Cytol.* 17, 299.

Hess, R., Barratt, D., and Gelzer, J. (1968). Immunofluorescent localization of β-corticotropin in the rat pituitary. *Experientia* 24, 584.

Hilderbrand, J. E., Rennels, E. G., and Finerty, J. C. (1957). Gonadotropic cells of rat hypophysis and their relation to hormone production. *Z. Zellforsch.* 46, 400.

Hinton, D. M., Kavanagh, W. G., Petrali, J. P., Lenz, D. E., Meyer, H. G., and Sternberger, L. A. (1971). Polymerization of peroxidase as possible cause for the unique composition and stability of soluble peroxidase-antiperoxidase complex in the unlabeled antibody peroxidase method. *J. Histochem. Cytochem.* **19**, 710.

Ishizaka, K., Ishizaka, T., and Campbell, D. H. (1959). The biological activity of soluble antigen-antibody complexes. II. Physical properties of soluble complexes having skin irritation activity. *J. Exp. Med.* **109**, 127.

Ishizaka, K., Ishizaka, T., and Hornbrook, M. M. (1970). A unique rabbit immunoglobulin having homocytotropic antibody activity. *Immunochem.* **7**, 515.

Jerne, N. K., Nordin, A. A., and Henley, C. (1963). The Agar Block Technique for Recognizing Antibody-Producing Cells. In *Cell Bound Antibodies* (Amos, B., and Koprowski, H., eds.), Wistar Institute Press, Philadelphia.

Kaplan, M. E., and Kabat, D. A. (1966). Studies on human antibodies. IV. Purification and properties of anti-A and anti-B obtained by absorption and elution from insoluble blood group substances. *J. Exp. Med.* **123**, 1061.

Kawarai, Y., and Nakane, P. K. (1970). Localization of tissue antigens on the ultrathin sections with peroxidase labeled antibody method. *J. Histochem. Cytochem.* **18**, 161.

Kendall, P. A. (1965). Labeling of thiolated antibody with mercury for electron microscopy. *Biochim. Biophys. Acta* **97**, 179.

Kuettner, K. E., Eisenstein, R., Soble, L. W., and Arsenis, C. (1971). Lysozyme in epiphyseal cartilage. IV. Embryonic chick cartilage lysozyme: its localization and partial characterization. *J. Cell Biol.* **49**, 450.

Kuhlmann, W. D., and Avrameas, S. (1971). Glucose oxidase as an antigen marker for light and electron microscopic studies. *J. Histochem. Cytochem.* **19**, 361.

Lafferty, K. G., and Oerteliss, J. (1963). The interaction between virus and antibody. III. Examination of virus-antibody complexes with the electron microscope. *Virology* **21**, 91.

Laguens, R., and Segal, A. (1969). Experimental autologous immune complex nephritis: an electron microscope and immunohistochemical study. *Exp. Mol. Path.* **11**, 89.

Leznoff, A., Fishman, J., Talbot, M., McGarry, E. E., Beck, J. C., and Rose, B. (1962). The cytological localization of ACTH in the human pituitary. *J. Clin. Invest.* **41**, 1720.

McShan, W. H., and Hartley, M. W. (1965). Production, storage, and release of interior pituitary hormones. *Ergeb. Physiol.* **56**, 264.

Mannik, M., Arend, W. P., Hall, A. P., and Gilliland, B. C. (1971). Studies on antigen antibody complexes. I. Elimination of soluble complexes from rabbit circulation. *J. Exp. Med.* **133**, 713.

Marshall, Jr., J. M. (1951). Localization of adrenocorticotropic hormone by histochemical immunochemical methods. *J. Exp. Med.* **94**, 21.

Mason, T. E., Phifer, R. F., Spicer, S. S., Swallow, R. A., and Dreskin, R. B. (1969). An immunoglobulin-enzyme bridge method for localizing tissue antigen. *J. Histochem. Cytochem.* **17**, 563.

Mayer, M. M. (1970). Highlights of complement research during the past twenty-five years. *Immunochem.* **7**, 485.

von Mayersbach, H. (1966). Immunohistologic Methods in Histochemistry. In *Handbuch der Histochemie*, Vol. 1, *General Methodology*, 2d part (Graumann, W., and Neuman, K., eds.). Gustave Fischer Verlag, Stuttgart.

von Mayersbach, H. (1967). The scope of nonspecific staining. *Acta Histochem.* Suppl. **7**, 271.

Minden, P., Grey, H. M., and Farr, R. S. (1967). False positive radioimmunoautography lines associated with immunoglobulins from normal sera. *J. Immunol.* **99**, 304.

Modesto, R. R., and Pesce, A. J. (1971). The reaction of 4,4'-difluoro-3,3'dinitrodiphenyl sulfone with ∝-globulin and horseradish peroxidase. *Biochem. Biophys. Acta* **229**, 384.

Moriarty, G. C., and Halmi, N. S. (1972). Electron microscopic study of the adrenocorticotropin-producing cell with the use of unlabeled antibody and the soluble peroxidase-antiperoxidase complex. *J. Histochem. Cytochem.* **20**, 590.

Nakane, P. K. (1968). Simultaneous localization of multiple tissue antigens using the peroxidase labeled antibody method: a study on pituitary glands of the rat. *J. Histochem. Cytochem.* **16**, 557.

Nakane, P. K. (1970). Classification of anterior pituitary cell types with immunoenzyme histochemistry. *J. Histochem. Cytochem.* **18**, 9.

Nakane, P. K., and Pierce, Jr., G. B. (1966). Enzyme-labeled antibodies: preparation and application for the localization of antigens. *J. Histochem. Cytochem.* **14**, 929.

Nakane, P. K., and Pierce, Jr., G. B. (1967). Enzyme-labeled antibodies for the light and electron microscopic localization of tissue antigen. *J. Cell Biol.* **33**, 307.

Novikoff, A. B., and Goldfischer, S. (1969). Visualization of peroxisomes (microbodies) and mitochondria with diaminobenzidine. *J. Histochem. Cytochem.* **17**, 675.

Ornstein, L. (1966). Discussion on enzyme-labeled antibodies for light and electron microscopic localization of antigens. *J. Histochem. Cytochem.* **14**, 790.

Paul, W. E., Katz, D. H., Goidl, E. A., and Benacerra, B. (1970). Carrier function in anti-hapten immune responses. II. Specific properties of carrier cells capable of enhancing anti-hapten antibody responses. *J. Exp. Med.* **132**, 283.

Phifer, R. F. (1970). Personal communication to Dr. Hardy.

Phifer, R. F., and Spicer, S. S. (1970). Immunohistologic and immunopathologic demonstration of adrenocorticotropic hormone in the pars intermedia of the adenohypophysis. *Lab. Invest.* **23**, 543.

Phifer, R. F., Spicer, S. S., and Orth, D. N. (1970). Specific demonstration of the human hypophyseal cells which produce adrenocorticotropic hormone. *J. Clin. Endocrin. Metab.* **31**, 347.

Pierce, Jr., G. B., and Nakane, P. K. (1967). Antigens of epithelial basement membrane of mouse, rat, and man: A study utilizing enzyme-labeled antibody. *Lab. Invest.* **17**, 499.

Pierce, Jr., G. B., Ram, J. S., and Midgley, A. R. (1964). Labeled antibodies in electron microscopy. *Int. Rev. Exp. Path.* **3**, 1.

Porter, R. R. (1959). The hydrolysis of rabbit gamma globulin antibodies with crystalline papain. *Biochem. J.* **73**, 119.

Pressman, D., Stelos, P., and Grossberg, A. (1961). Retention of rabbit antibody during acetylation. *J. Immunol.* **86**, 452.

Quiocho, F. A., and Richards, F. M. (1964). Intermolecular crosslinking of a protein in the crystalline state: carboxypeptidase-A. *Proc. Nat. Acad. Sci.* **52**, 833.

Ram, J. S., Nakane, P. K., Rawlinson, E. G., and Pierce, Jr., G. B. (1966). Enzyme-labeled antibodies for ultrastructural studies. *Fed. Proc.* **25**, 732.

Richards, F. M., and Knowles, J. K. (1968). Glutaraldehyde as a protein crosslinking reagent. *J. Mol. Biol.* **37**, 231.

Sabatini, D. C., Bensch, K., and Barrnett, R. J. (1963). The preservation of cellular ultrastructure and enzymatic activity by aldehyde fixation. *J. Cell Biol.* **17**, 19.

Schiff, R., Krieg, R. J., and Hunter, L. (1970). Localization by peroxidase-labeled antibodies of bovine chymotrypsinogen. *J. Histochem. Cytochem.* **18**, 195.

Schlossman, S. F., and Kabat, D. A. (1962). Specific fractionation of a population of antidextran molecules with combining sites of various sizes. *J. Exp. Med.* **116**, 535.

Schreiner, E., and Wolff, K. (1970). Systemic *Lupus erythematosus:* electron microscopic localization of *in vivo* bound globulins at the dermal-epidermal junction. *J. Invest. Dermatol.* **55**, 325.

Seligman, A. M. (1971). Unreliability of ultrastructural demonstration of enzymes which depend upon hydrogen peroxide production from action on their substrates. *J. Histochem. Cytochem.* **19**, 809.

Seligman, A. M., Wasserkrug, H. L., and Hanker, J. S. (1966). A new staining method for enhancing contrast of lipid-containing membranes and droplets in osmium-fixed tissue with osmiophilic thiocarbohydrazide. *J. Cell Biol.* **30**, 424.

Singer, S. J. (1959). Preparation of an electron dense antibody conjugate. *Nature* **183**, 1, 523.

Singer, S. J., and Campbell, D. H. (1952). Physical chemical studies of soluble antigen-antibody complexes. I. The valence of precipitating rabbit antibody. *J. Amer. Chem. Soc.* **74**, 1794.

Singer, S. J., and Campbell, D. H. (1953). Physical chemical studies of soluble antigen-antibody complexes. II. Equilibrium properties. *J. Amer. Chem. Soc.* **75**, 5577.

Singer, S. J., and Campbell, D. H. (1955). Physical chemical studies of soluble antigen-antibody complexes. IV. The effect of pH on the reaction between BSA and its rabbit antibodies. *J. Amer. Chem. Soc.* **77**, 3504.

Singer, S. J., and Schick, A. F. (1961). The properties of specific stains for electron microscopy prepared by conjugation of antibody molecules with ferritin. *J. Biophys. Biochem. Cytol.* **9**, 519.

Sternberger, L. A. (1967). Electron microscopic immunocytochemistry: A review. *J. Histochem. Cytochem.* **15**, 139.

Sternberger, L. A. (1969). Some new developments in immunocytochemistry. *Mikroskopie* **25**, 346.

Sternberger, L. A. (1972). The Unlabeled Antibody Peroxidase and the Quantitative Immunouranium Methods in Light and Electron Immunohistochemistry. In *Techniques of Biophysical and Biochemical Morphology*, Vol. I, (D. Glick and R. M. Rosenbaum, eds.), Wiley-Interscience, New York.

Sternberger, L. A. Cuculis, J. J., Meyer, H. G., and Hoy, N. J. (1961). The precipitation of nonprecipitating and redissolved precipitating antigen-antibody complex by ammonium sulfate, anti-antibody, and alkali treatment. *Fed. Proc.* **20**, 20.

Sternberger, L. A., Donati, E. J., Hanker, J. S., and Seligman, A. M. (1966a). Immunodiazothioether osmium tetroxide (immuno-DTO) technique for staining embedded antigen in electron microscopy. *Exp. Mol. Path. Suppl.* **3**, 36.

Sternberger, L. A., Donati, E. J., Petrali, J. P., Hanker, J. S., and Seligman, A. M. (1966b). Method for enhancement of electron microscopic visualization of embedded antigen by bridging osmium to uranium antibody with thiocarbo-hydrazide. *J. Histochem. Cytochem.* **14**, 711.

Sternberger, L. A., Hardy, Jr., P. H., Cuculis, J. J., and Meyer, H. G. (1970a). The unlabeled antibody enzyme method of immunohistochemistry: Preparation and properties of soluble antigen-antibody complex (horseradish peroxidase-antihorseradish peroxidase) and its use in identification of spirocheetes. *J. Histochem. Cytochem.* **18**, 315.

Sternberger, L. A., Osserman, E. F., and Seligman, A. M. (1970b). Lysozyme and fibrinogen in normal and leukemic blood cells: a quantitative electron immunocytochemical study. *Johns Hopkins Med. J.* **126**, 188.

Sternberger, L. A., Hinton, D. M., Petrali, J. P., Meyer, H. G., and Cuculis, J. J. (1971). Approach to quantitative staining histochemistry. *J. Histochem. Cytochem.* **19**, 710.

Terry, W. D., and Fahey, J. L. (1964). Subclasses of human $_2$ globulin based on differences in heavy polypeptide chains. *Science* **146**, 400.

Ubertini, T., Wilkie, B. L., and Noronha, F. (1971). Use of horseradish peroxidase-labeled antibody for light and electron microscopic localization of reovirus antigen. *Appl. Microbiol.* **21**, 534.

Van Furth, R., Hirsch, J. G., and Fedorko, M. E. (1970). Morphology and peroxidase cytochemistry of mouse promonocytes, monocytes, and macrophages. *J. Exp. Med.* **132**, 794.

Vogt, A., and Kopp, R. (1965a). Loss of specific agglutinating activity of purified ferritin-conjugated antibodies. *Nature* **202**, 1350.

Vogt, A., and Kopp, R. (1965b). Specific activity of ferritin-labeled antibodies. *Z. Bakt.* **198**, 270.

Vreeland, V. (1970). Localization of a cell wall polysaccharide in a brown alga with labeled antibody. *J. Histochem. Cytochem.* **18**, 371.

Weliky, N., and Weetall, R. V. (1965). The chemistry and use of cellulose derivatives for the study of biological systems. *Immunochemistry* **2**, 293.

Weliky, N., Weetall, R. V., Gilden, R. V., and Campbell, D. H. (1964). The synthesis and use of some insoluble immunologically specific adsorbents. *Immunochemistry* **1**, 219.

White, J. G. (1970). The interplatelet zone. *Amer. J. Pathol.* **58**, 19.

Wide, L., Axen, R., and Porath, J. (1967). Radioimmuno-sorbent assay for proteins: Chemical coupling of antibodies to insoluble dextran. *Immunochemistry* **4**, 381.

Wolff, K., and Schreiner, E. (1970). Immunohistochemical studies with an enzyme-immunoglobulin conjugate: Detection in *Lupus erythematosus, Pemphigus vulgaris,* pemphigoid of immunoglobulin bound *in vivo. Arch. Klin. Exp. Derm.* **238**, 381.

Yagi, Y., Maier, P., Pressman, D., Arbesman, C. E., and Reisman, R. E. (1963). The presence of the ragweed binding antibodies in the β_{2A}-, β_{2M}- and α-globulins of the sensitive individuals. *J. Immunol.* **91**, 83.

Zacks, S. I., and Scheff, M. F. (1968). Tetanus toxin: Fine structural localization of binding sites in striated muscle. *Science* **159**, 643.

Zamboni, L., and DeMartino, C. (1967). Buffered picric acid formaldehyde: a new rapid fixative for electron microscopy. *J. Cell Biol.* **35**, 148a.

Zeromski, J., Perlmann, P., Lagercrantz, R., and Gustafsson, B. E. (1970). Immunological studies with peroxidase-conjugated antibodies. *Clin. Exp. Immunol.* **7**, 463.

Author Index

Subject Index

203

Science

Date Due

APR 1 2 1976		
SEP 1 6 1976		
APR 1 1 1977		
DEC 1 6 1979		
MAY 2 6 1982		
NOV 1 2 1982		
AUG 3 1 1988		
		UML 735